FUNDAMENTALS OF TELEVISION

theory and service

Bernard Hartman

Quasar Electronics Corporation

Charles E. Merrill Publishing Company
A Bell & Howell Company
Columbus, Ohio

Published by
Charles E. Merrill Publishing Company
A Bell & Howell Company
Columbus, Ohio 43216

International Standard Book Number: 0–675–08745–7

Library of Congress Catalog Card Number: 74–16860

1 2 3 4 5 6 7 8 9 10—81 80 79 78 77 76 75

Printed in the United States of America

FUNDAMENTALS OF TELEVISION

theory and service

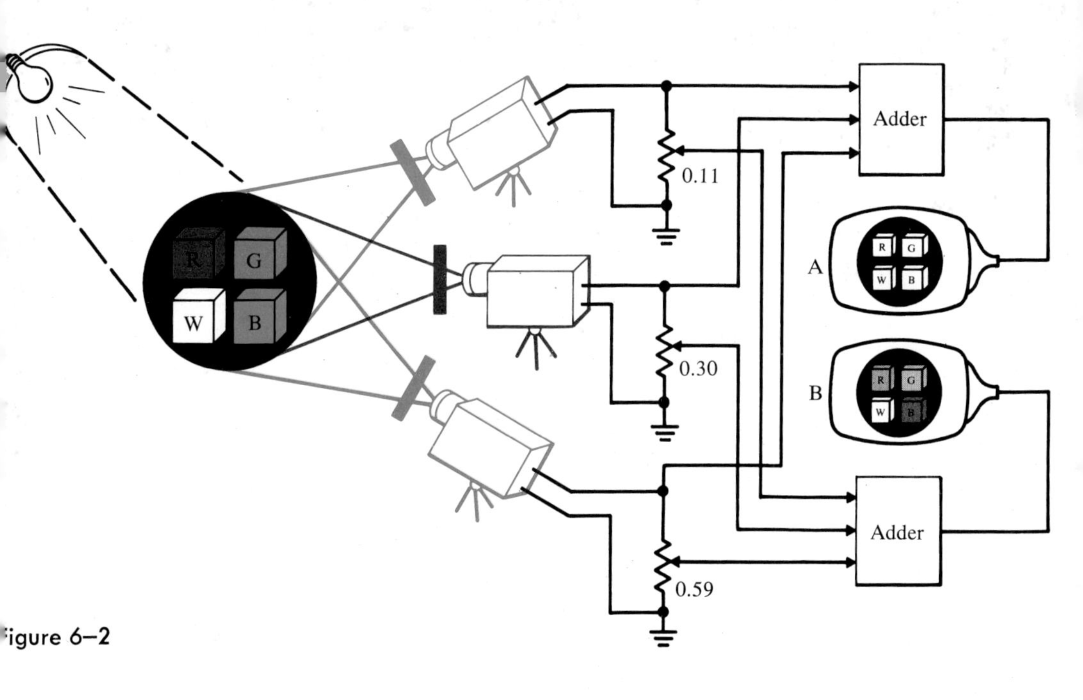

Figure 6–2

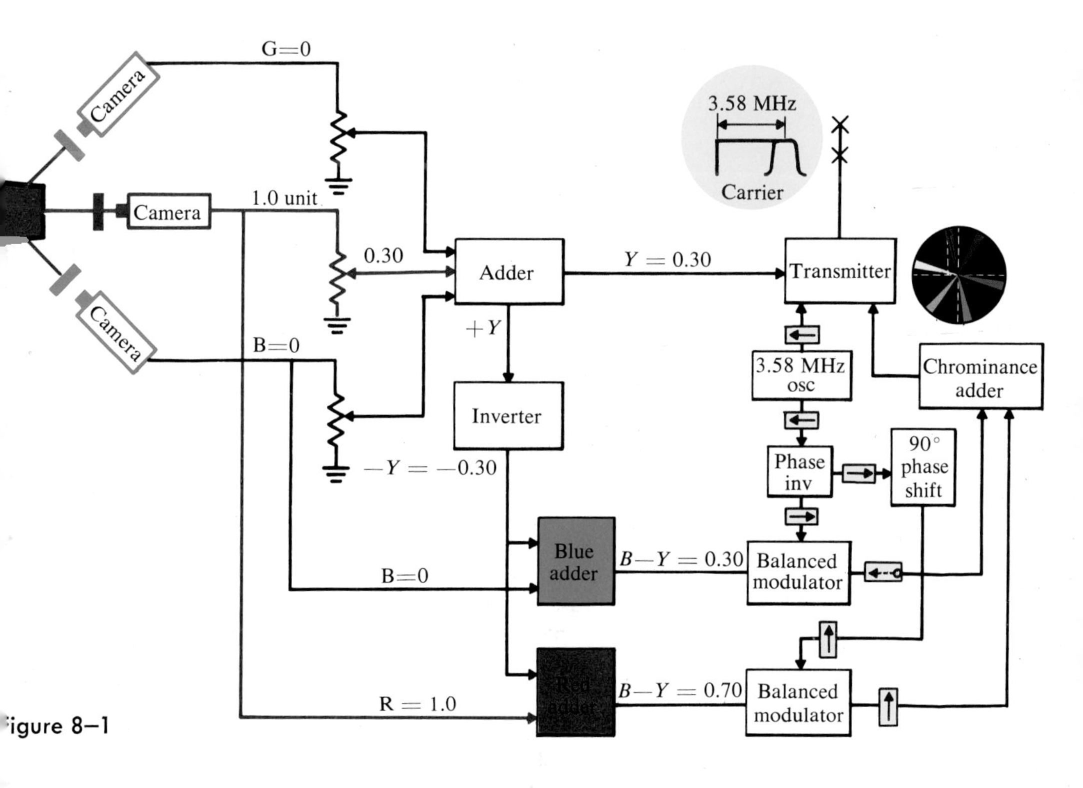

Figure 8–1

To the memory of
Jack and Fanny Hartman
and Herman Weinstock

preface

On October 30, 1946, Radio Corporation of America publicly demonstrated its first all-electronic means of producing color pictures. In the short time since, color television has become a reality.

This text is the result of a need for a more uniform approach to both theory and service. The future of qualified servicemen is dependent upon an understanding and appreciation of both of these. The text is designed for students of technology and is planned for a semester course.

Chapters 3,4,6,8, and 11 are inserted as a review in the following subject matter: sine-wave principles, black-and-white television, modulation, magnetism, and transistor theory. An understanding of these basics is essential to help the student comprehend the material that follows. Only those basics that are necessary have been put to print. It is not implied that all who use this book have deficiencies in these basics, nor is it implied that only the basics described are necessary for the qualified service technician.

Primarily, the method used throughout this book is the systems or model approach. Some circuit analysis and description was necessary to prepare for the chapter on convergence and panel analysis. The service-bench technician who desires to repair panels should have a well-rounded technical background.

Part I introduces the reader to color fundamentals, and Part II introduces the reader to the new and exciting side of in-home and shop service.

The vacuum tube is being phased out rapidly and the transistor is taking its place. Integrated circuits are replacing many of the functions of the transistors. The three-gun color picture tubes have progressed from 70° deflection to 90° deflection. Eventually they will achieve 110° deflection. With each increase of deflection angle, the tube length is reduced.

Receivers are becoming smaller in size due to the technological improvements. The color receiver can be styled attractively because the "big-box" concept is no longer necessary.

Color-tube phosphors are being improved to reproduce more vivid colors. The black-around and matrix tubes provide the viewer with better contrast ratios and brighter pictures.

It is difficult for one to imagine that more progress can be made. Have we reached the end? Let us speculate at possible advances that may be made in the future.

Three-dimensional television may be developed for more realism. In 1947, 35mm slides were in vogue. There were experiments with screens utilizing thousands of fine, vertical, cylindrical columns on which two separated pictures were projected so that each eye of an observer would see slightly different images. True, there were distortion problems, but they were not unsolvable. Who has not seen and heard stereophonic cinemascope movies?

Two-color systems have been demonstrated, but still leave a lot to be desired. Certainly this is not the end to this approach. The one-gun tube has its drawback because of radiation. Surely, this problem will be overcome if economics warrant it.

How about the flat tube on the wall with the customer controls packaged in a styled cabinet no larger than a cigar box? This must excite the interior decorator and the home architect.

The potential of television is limitless, as long as man continues to explore. To fulfill this potential, man's thirst for knowledge must never diminish.

acknowledgements

The author is deeply indebted to Quasar Electronics Corporation (formerly Motorola Consumer Products and referred to as Motorola throughout this text), but specifically to Larry Wren and Bob Nielson. These gentlemen gave me the opportunity to find that teaching is a rewarding avocation. The material for this book evolved from a collection of material accumulated through many years of teaching.

Special mention should be made of the help received from the technical writing staff of Motorola for so much of what is forthcoming in this book, specifically chapters 9 and 10.*

Wally Bass of Quasar Electronics for his concentrated help with chapters 11 to 15.

Appreciation is also given to the following:

The Navy for their permission to use material for chapter 3.

---

* Motorola "Fundamentals of Color TV Manual 68P65110A28-0"

McGraw-Hill Book Company for their permission to use material from *Elements of Electronics* by Hickey and Villines and *Color Television Engineering* by Wentworth for chapter 1 and chapter 6.

Coyne American Institute for permission to use definitions from *Coyne Technical Dictionary*. These definitions can be found in the glossary.

Professors Donald Beaty, Louis Gross, Chester Jur, and Paul Muxlow for their constructive reactions and thought-provoking suggestions.

This whole project would never have happened were it not for W. R. Hutton, Jr., and Joanne Morelli Harris with her patience and steadfast position for perfection.

Finally, to my wife, Bea, and son and daughter, Alan and Fern, for their patience, help, and encouragement.

contents

13 functional diagram 171

14 in-home service 185

FUNDAMENTALS OF TELEVISION

theory and service

part one

FUNDAMENTALS

introduction

As we view our natural surroundings, it is obvious that there is more to see than the different shades of gray from black to white. Describe a tree, grass, or the complexion of a person without using color as your adjective. As in photography, television's transition to color was a must to give man a realistic picture. Developing technology made the must a reality.

As we will learn in the following chapters, lab experiments determined that three colors could be used to reproduce a color scene. This new knowledge won only half the battle. Next came the problem of transmission. The signals necessary to convey color information had to be transmitted to a receiver designed to pick up color signals while not making the monochrome (black-and-white) receivers that were in use at the time obsolete.

Columbia Broadcasting System was at first thought to be the best practical approach. This system used a rotating color disc placed in front of the television camera tube. The speed of the color disc was synchronized with the electron-beam scanning of the camera tube. It contained three primary-color filters—red, green, and blue. To scan an image completely all colors required six scannings with the electron beam. At the receiver a color disc was again needed and was synchronized so that the color filter was in position at the precise moment that the particular color was being transmitted. The disadvantage of this system was the reduction of light due to the filters. An increase in scanning rate was necessary to eliminate flicker and a disc would add a mechanical element to the receiver. More importantly, this system did not conform with the National Television System Committee's objective of achieving compatibility with monochrome transmission.

The National Television System Committee consisted of representatives from about one hundred concerns, all of whom had some interest in the setting of color broadcasting standards. Their task was to set tentative specifications and field test them. The committee's final suggestions were filed with the FCC for approval on July 23, 1953. On December 17, 1953, the FCC approved the signal specifications proposed by the committee. They were adopted as technical standards for commercial color television broadcasting in the United States and will be discussed further in chapter 6.

1 nature of light

Characteristics of Light

Color is a form of light. To understand color it is first necessary to understand the nature of light.

Light is considered to be electromagnetic energy similar to radio and television signals. It obeys similar laws. All electromagnetic radiations with wavelengths between 0.00004 and 0.00007 centimeter appear as light. Because these decimal fractions are rather cumbersome to work with, a smaller unit, the *millimicron,* is used to express wavelength of light. A millimicron is one thousandth of one millionth of one meter. Visible light can be expressed as having wavelengths between 400 and 700 millimicrons.

Figure 1–1 is a graphic presentation of the total electromagnetic radiation spectrum showing the location of the wavelengths we call light.

At the left-hand side, low-frequency audio vibrations are shown. Note that the wavelengths get shorter at the right (indicating frequency increase), spanning the radio, television, and microwave regions. Next shown is the infrared area, where heat radiation takes place. Then, the frequencies which represent visible light are illustrated. The extreme right-hand portion is the domain of x rays, gamma rays, and cosmic rays.

If we examine the visible light portion closely, we see that each wavelength (or frequency) of visible light corresponds to a definite color.

Electromagnetic Radiation Spectrum

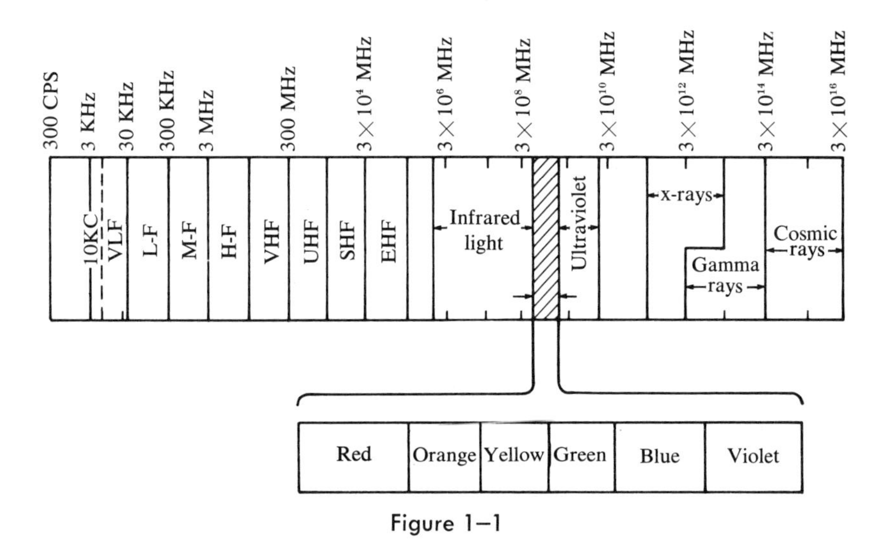

Figure 1–1

White light contains all of these wavelengths. We can demonstrate that white light is composed of all colors by use of a glass prism (Figure 1–2 —see also front endpapers).

When a ray of white light enters the glass of the prism, it passes from one transmission medium to another. This causes the light ray to bend. The amount of bending is governed by the refractive index of the glass and the wavelength of the light. Since the refractive index is different for each wavelength of light, the long wavelengths, or the red colors, do not bend as much as the shorter wavelengths, or the blue colors. The prism separates all components of light by wavelength and indicates that white light is made up of all wavelengths of color light.

It is possible to reverse the process and create white light by recombining all the colors in the spectrum of Figure 1–2. The rainbow reproduces the gamut of colors that we see in Figure 1–2. They range from red to violet. The rainbow is a phenomenon that is created by the selective dispersion of sunlight. The raindrop (water) is the medium which causes the separation of colors, as is the prism in Figure 1–2. Since sunlight is considered white, it can be concluded that white light is the result of all wavelengths of light presented to the eye simultaneously. The human color vision theory is evolved from these principles.

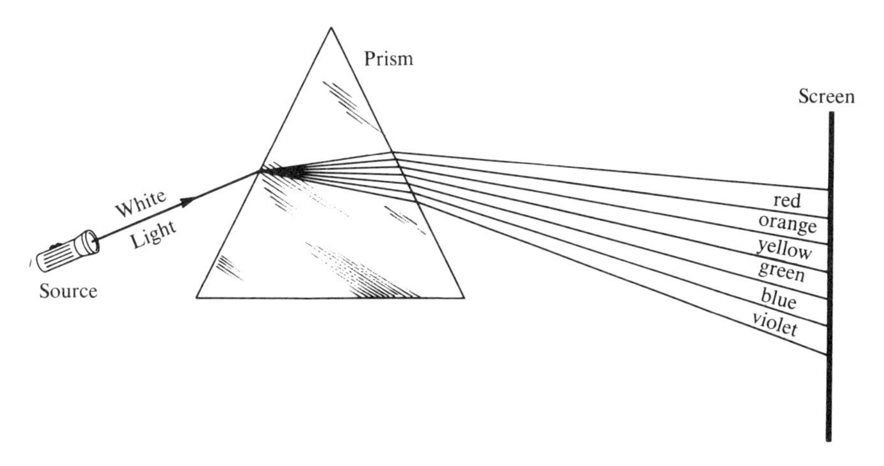

Figure 1–2

Characteristics of the Eye

At this time it would be helpful to review the physiological processes of the human eye. (See sketch of human eye shown in Figure 1–3.)* The human eye is a little like a camera in that it consists of a lens system at the front that focuses an inverted image of an external scene onto a light-sensitive surface, or retina, at the back. The camera and the eye differ in the way they adjust focus, correct aberrations, and adjust sensitivity.

Both the camera and the eye have variable focus abilities. In the camera, correct focus is achieved by changing distance from lens to film. In the eye, the actual focal length of the lens itself is changed by means of muscles which control the curvature of the lens surfaces.

The lens of the eye causes color aberration because the short-wavelength (blue) components of light come to focus at a point closer to the

---

* This section adapted from *Color Television Engineering* by John W. Wentworth. Copyright 1955 by McGraw-Hill Book Company. Used with permission of McGraw-Hill Book Company.

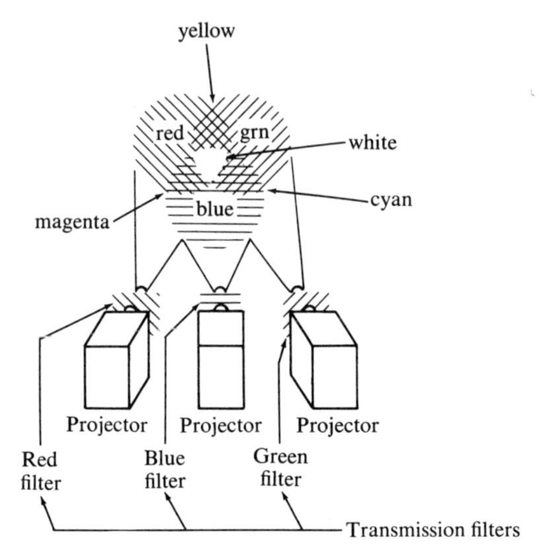

Figure 1–3

lens than the long-wavelength (red) components. The correction for the color aberration in the human eye is never complete; the blue light may be attenuated, but it is never brought into sharp focus. The result is that the resolving power of the eye is very low in blue areas.

The retina is the film of the eye. It consists of light-sensitive cells, namely, rods and cones. The rods are the only cells responsible for night vision. They are highly sensitive but do not provide color vision. The cone cells function only when the illumination level is above approximately 0.1 foot-candle. These are the color-sensitive cells.

The "blind spot" of the eye is the point of contact with the optical nerves located at the rear of the eye. At this point there are no rods or cones. There is another spot, approximately one-half millimeter in diameter, near the center of the retina where rods are completely lacking, but where the cones are numerous. This area is called the fovea. The absence of rod cells in the fovea makes it difficult for us to see objects in very low illumination. The cones are composed of cells with overlapping curves such that some are more sensitive to blue, others to red, and still others, to green.

Human Vision and Television Compatibility

The structure and workings of the human eye influenced the design of a compatible color television system in two ways. First, design had

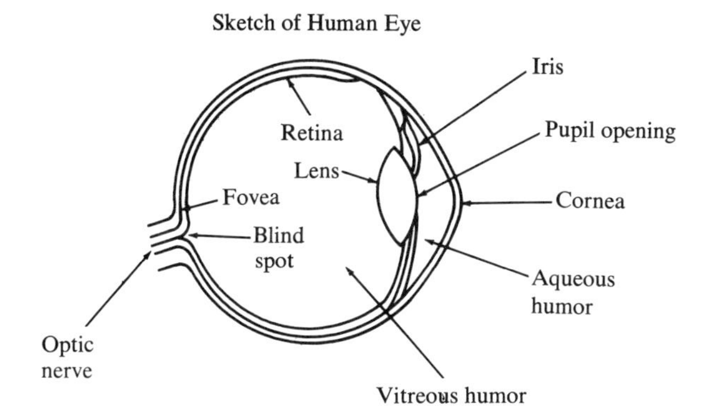

Figure 1–4

From Color Television Engineering *by John W. Wentworth. Copyright 1955 by McGraw-Hill Book Company. Used with permission of McGraw-Hill Book Company.*

to take into account the fact that we do not see all colors in the same proportions. Figure 1–5 illustrates four 100-watt light bulbs, each a different color. Each bulb radiates the same amount of light energy, but the eye "sees" these as different amounts of light output.

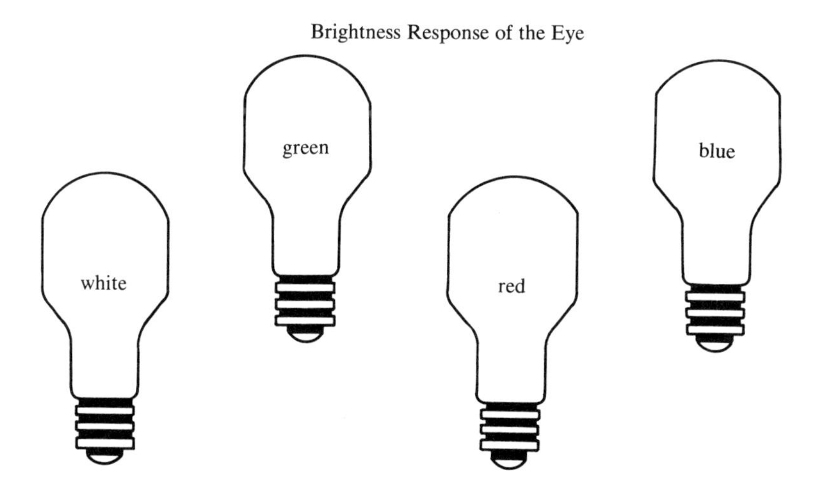

Figure 1–5

The white bulb appears the brightest. The green is 59% as bright, the red 30% as bright, and the blue only 11% as bright as the white bulb. This can be plotted graphically as shown in Figure 1–6, which shows a bandpass response curve of the eye which relates the amount of stimulation at the optic nerve to the wavelength of light being observed. The graph shows us that the maximum response or stimulation to the eye takes place in the green-yellow region, which is about 555 millimicrons wavelength. There is a lesser response in the blue and red regions, making these two colors appear less bright than green.

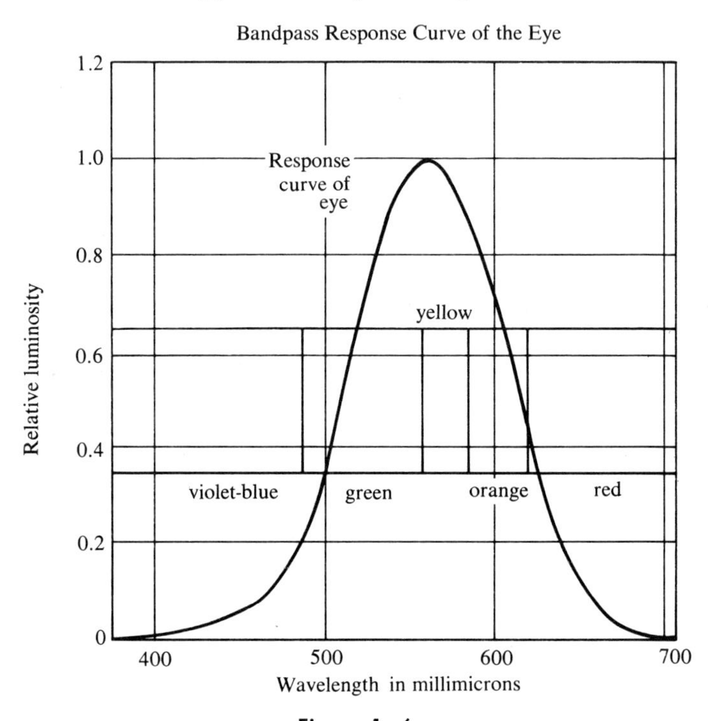

Figure 1–6

The second characteristic of the eye important to the principle of compatibility is its ability to resolve fine detail. This ability varies with the color observed. Basically, color identification becomes more difficult as the size of the color area is reduced. This means that color is seen only in large areas and not as fine detail. As colored objects are reduced in size, four things happen in succession. First, blues become indistinguishable from grays of equivalent brightness. Second, yellow becomes indistinguishable from grays. Third, with still further decrease in size, reds merge with grays. Finally, blue-greens also become indistinguishable from gray. The eye perceives brightness detail in areas 5 to 10 times smaller than it perceives color detail. As color dots become very small, they loose their identification as dots in terms of shades of gray only.

We must understand that in our discussion of color mixture we are dealing with *additive* colors (light), not *subtractive* colors (mixtures of pigments).

The primary colors in the subtractive system, as used in modern printing, paints, and slide transparencies, are yellow, cyan (blue-green), and magenta. When all three are added, we get black.

Red, green, and blue are the primary colors in the additive system. When they are added, the net result is white. Later we will see that a mixture of unequal proportion will yield white. We will also observe that not all the possible colors are reproduced by the addition of these primaries.

Brightness, Hue, and Saturation

There are three characteristics of light by which we can define any source of light that we can observe or produce.

The first of these characteristics is *brightness*. Brightness is defined as the amount of light, or the amount of energy, reaching the eye from the scene, independent of all other considerations. It is brightness that determines the shades of gray that we perceive on a black-and-white television, for it is brightness, not hue, that is translated into grays by the eye. Figure 1–7 shows two light bulbs both giving off a white light. The 100-watt bulb gives roughly four times more brightness, or energy output, than the 25-watt bulb.

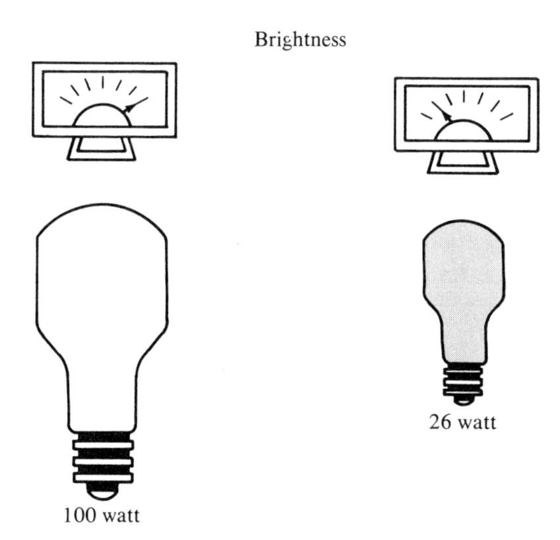

Figure 1–7

The second characteristic of light in which we are interested is *hue*. Hue is defined as the wavelength of the color of the light, regardless of all other considerations. Figure 1–8 shows two red light bulbs, a 100-watt bulb and a 25-watt bulb. The color, or the hue, emanating from these bulbs is identical even though the 100-watt bulb is putting out roughly four times as much energy as the 25-watt bulb. What we must note here is that, regardless of brightness, they are giving off the same color, or the same wavelength, of light.

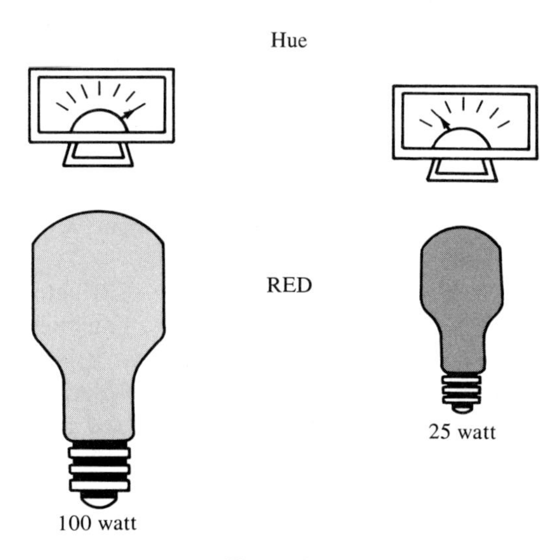

Figure 1–8

The third characteristic of light is *saturation*. Saturation is defined either as the absence of dilution with white light or as the "pureness" of the color. Figure 1–9 shows a fully saturated red beam of light being intersected by a pure white beam of light. Where the two beams intersect, a color we recognize as pink is created. Pink is a desaturated red color. All pastel shades are desaturated pure colors.

A pure color is one that is 100% saturated. As we dilute the color by adding white light, the percentage of saturation decreases. For example, if we dilute 100% saturated color with 80% white light, the result will yield a 20% saturated color (or a pure color that is 80% desaturated).

To summarize, in order to transmit a color picture, we must transmit brightness, hue, and saturation information. But to make the transmissions compatible with existing black-and-white receivers we must also take into account the characteristics of the eye. This means that the black-and-white receiver should reproduce the various colors as shades of gray (brightness), and there should be no deterioration of detail in the black-and-white picture.

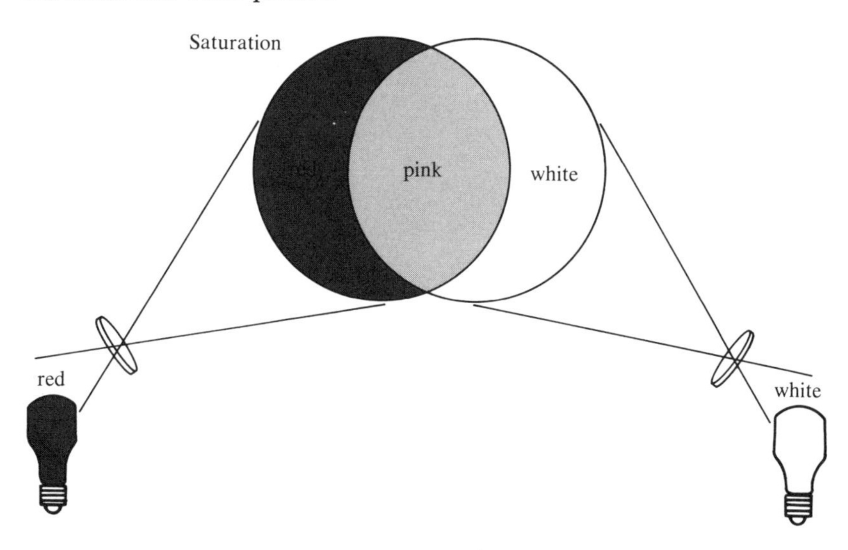

Figure 1–9

questions

1. The refraction index is different for each wavelength.
   True _____ False _____
2. The long wavelengths of red bend (more), (less) than the short wavelengths of blue.
3. White is a combination of what colors in the additive process?
4. The (rods), (cones) of the eye are sensitive to color.
5. Of the three primary colors, which would appear the brightest to the eye?
6. Yellow appears brightest because it falls in the center of the luminosity response of the eye. True _____ False _____
7. Complete the following:
   Red light plus green light = __________ light
   Red light plus blue light = __________ light
   Blue light plus green light = __________ light
8. The three characteristics of light are hue, brightness, and __________.
9. (Hue), (Saturation) describes the color that is to be transmitted.
10. Brightness information has to have greater detail than color information. True _____ False _____

2 black-and-white television

Radio is the science of transmitting sound by means of electromagnetic waves. Television is the science of transmitting both visual images and sound by means of electromagnetic waves.

Picture Standards

Due to a characteristic of the human eye known as persistence of vision, motion pictures are possible. Still images of a moving figure shown rapidly, at different increments of time, will appear to the eye as if in continuous motion. In motion pictures, still images are shown at a rate of 24 images (frames) per second with a shutter moving across each as it is projected. This process reproduces each picture twice, giving the film an effective rate of 48 frames per second.

In television, a fundamental rate of 30 frames per second are projected in order to relate to the frequency of power lines (vertical rate). In place of the shutter used in the motion picture projector, an electronic method called *interlaced scanning* is used. The picture is scanned twice. The eye, which cannot separate the fact that it is seeing 30 frames twice, effectively, sees 60 frames per second (Figure 2–2).

Scanning Process

A fine-detailed picture is composed of many elements. It is possible to transmit each element separately in an orderly sequence by a process called horizontal scanning. Figure 2–1 illustrates the scanning process. When the camera focuses on an object, it emits an electron beam that transverses (traces) from left to right from point *A* to point *B,* returning to point *C.* (Notice the similarity to the reading motion of your eyes.)

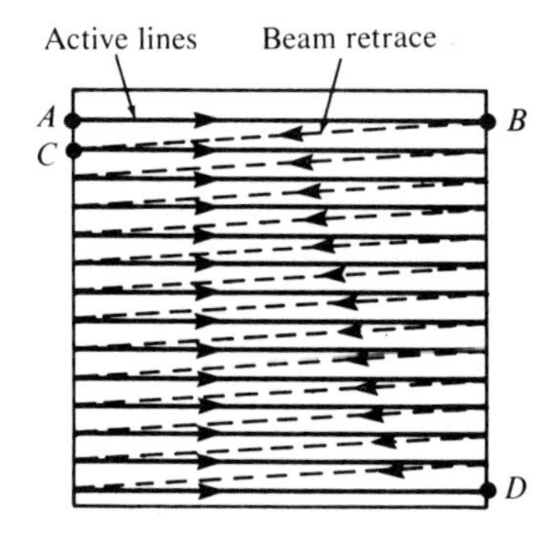

Figure 2–1

The returning beam is called *retrace.* Retrace occurs at a time when the camera tube is not allowed to receive an image. This is accomplished electronically and is known as *blanking.* Simply stated, the camera-tube gun is not permitted to emit electrons during the retrace portion of the scan. This sequence repeats until the end of the last line at point *D.* At this time the beam is again blanked and the process repeats.

Aspect Ratio

Each complete image is sent at a rate of 30 frames per second. To obtain the detail desired, a total of 525 horizontal lines are reproduced for a given picture. Since we divide the picture into two segments, or fields, 262½ lines exist in each field. It takes two fields for one complete picture (525 lines).

In Figure 2–2 we see that the first field starts at *A*, transverses horizontally to *B*, and is completed at point *D*. The second field starts at point *E* and ends at point *F*. Notice that the second field interlaces with the first, filling in the spaces between the scanning lines of the first field. The aspect ratio is 4 : 3. This is a ratio equal to picture width divided by picture height. The aspect ratio 4 : 3 is similar to that used for motion picture film standards.

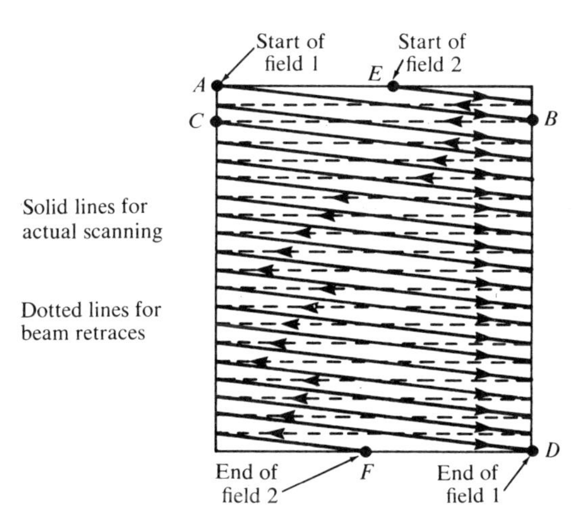

Figure 2–2

Composite Signal

The transmitted signal must include picture information (video), sound information (audio), and synchronizing information (vertical and horizontal). Figure 2–3 illustrates the composite signal for two lines of trace. One line of video information starts at point *A* and finishes at point *B*. At the end of one line of scan the blanking pulse is produced. This pulse cuts off the electron guns of the picture tube during retrace time. (Electron-gun structure will be explained in a later chapter.) The synchronizing pulses are placed on top of the blanking pulse. The synchronizing pulses are necessary to assure that the receiver and transmitter are scanning and reproducing picture elements at the same time.

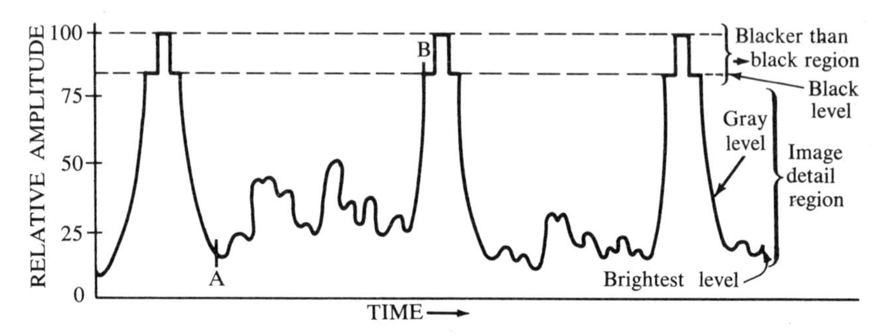

Figure 2–3

Sweep Rate

Figure 2–4 illustrates the makeup of the vertical and horizontal synchronizing pulses. It is necessary to synchronize the television horizontal and vertical sweeps. The composite signal amplitude-modulates the carrier frequency and the audio is frequency-modulated. Modulation will be discussed in greater detail in chapter 7.*

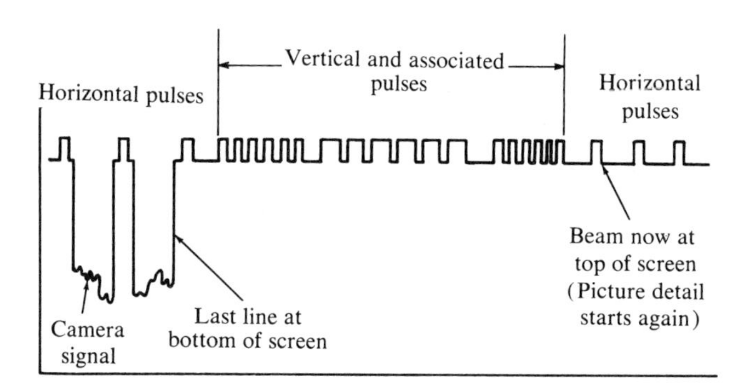

Figure 2–4

As previously stated, the more elements in a picture, the finer the detail. In one-thirtieth of a second, 525 lines are scanned. A total of 15,750 (30 x 525) lines are scanned in one second. To repeat, under present standards, 525 horizontal lines are scanned in one-thirtieth of a second or a total of 15,750 lines in one second. Employment of ap-

---

* Pages 18–21 adapted from *Television Simplified* by Milton S. Kiver and Milton Kaufman. Copyright 1973 by Litton Educational Publishing, Inc. Used with permission of Van Nostrand Reinhold Company.

proximately 508 elements per line results in the transmission of 8,000,000 picture elements per second. (This figure was determined by multiplying 508 x 15,750.) Thus we see how the rate of horizontal scanning is set. Because we use two fields per frame to obtain the total frequency bandwidth, we must divide these 8,000,000 elements by two, arriving at a bandwidth of approximately 4 MHz. It is compulsory for all the circuits in the video section to have the capability of passing this bandwidth. Since a bandwidth of 4 MHz is used, the total bandwidth per channel is 6 MHz; 2 MHz remain. The frequency-modulated audio requires a bandwidth of 50 kHz.

As can be seen by Figure 2–5, the remainder of the 6-MHz bandwidth, or 1.25 MHz, is used up by the vestigial side band.

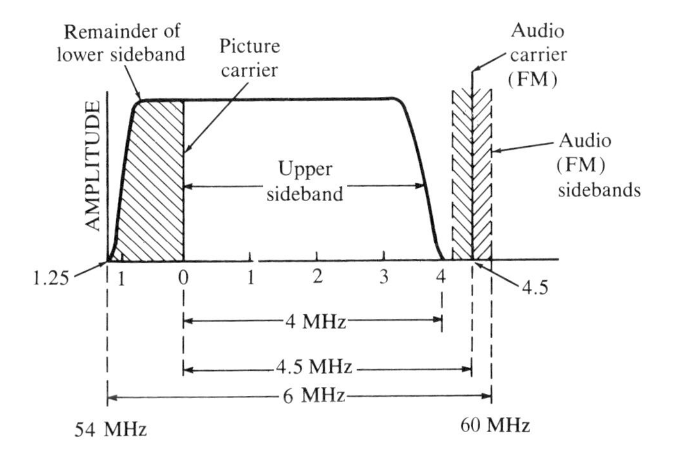

Figure 2–5

Now that the basic scanning process has been described, we move on to explanation of the television station transmitter and the television receiver.

Transmitter Block Diagram

Figure 2–6 is a simplified block diagram of a television station's transmitter. The picture viewed by the TV camera is translated into electrical impulses as the beam scans the surface of the camera tube. The electrical information is amplified by a video amplifier and fed to the modulation stages of the transmitter where it is superimposed onto the allocated frequency of the television station. The audio is frequency-modulated and is also superimposed onto the transmitted carrier. Scanning and synchronizing circuits are fed to the camera and transmitter.

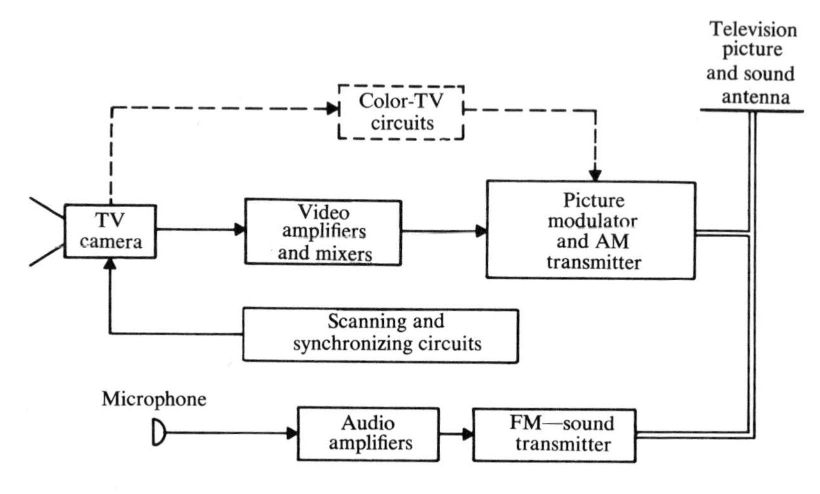

Figure 2–6

From Television Simplified *by Kiver and Kaufman.*
Copyright 1973 by Litton Educational Publishing, Inc.
Reprinted by permission of Van Nostrand Reinhold Company.

Receiver Block Diagram

Figure 2–7 is a simplified diagram of a television receiver. A tuner selects the channel frequencies desired (very high frequencies—vhf; ultrahigh frequencies—uhf). The channel frequencies are shown in Appendix C.

The output of the tuner is fed to the IF (intermediate frequency) stages. Although the tuner frequencies are different for each channel, the intermediate frequency is the same for all channels selected. The tuner converts (changes) all channel frequencies to the common intermediate frequency.

The outputs of the IF stages are fed to the video-detector stage. The function of the video detectors is to remove the video information, ranging from 30 Hz to 4.5 MHz, from the IF frequency. At this point the frequency-modulated audio carrier is also removed and forwarded to the sound-detector stage. The audio information will be recovered at the sound detector and amplified. The speaker will then convert the electrical impulses to mechanical vibrations which the ear will convert to the original sound information produced at the studio. The video information is recovered at the video detector, is amplified, and fed to the receiver picture tube. The picture tube converts the electrical impulses to picture information on its face.

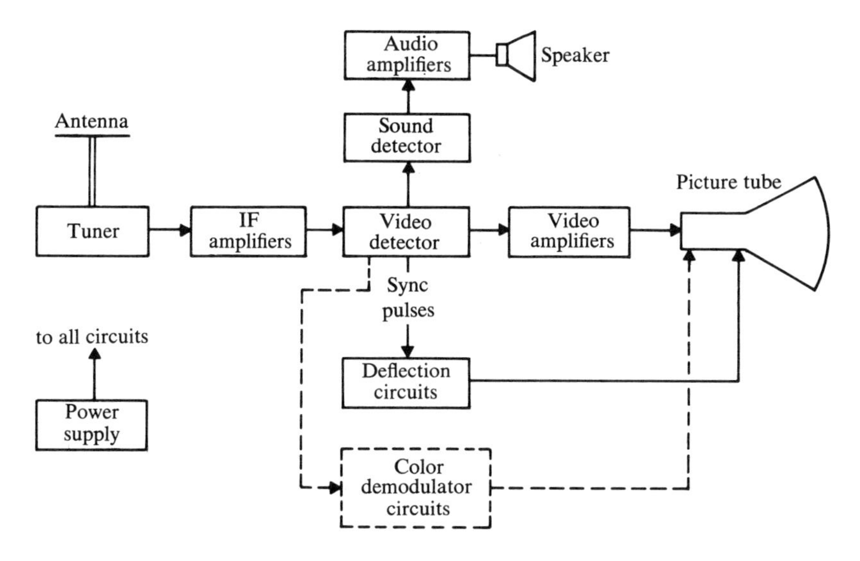

Figure 2–7

From Television Simplified *by Kiver and Kaufman.*
Copyright 1973 by Litton Educational Publishing, Inc.
Reprinted by permission of Van Nostrand Reinhold Company.

As we will see in chapters to follow, the black-and-white (monochrome) television block diagram and the color television diagram are basically the same. Color block functions are the only addition. The synchronizing frequencies are also altered slightly. The reasons for this change will be explained in the following chapters.

This chapter presents only an overview of black-and-white television principles. This is so by design, not oversight. In chapters covering the color television receiver, a more detailed description of the black-and-white sections of the receiver will be given.

questions

1. To achieve a fine-detailed picture (many), (few) elements must be reproduced.
2. Scanning is accomplished by movement of the beam from ________ to right.
3. Movement of the electron beam from left to right is known as (trace), (retrace).
4. The return of the electron beam from right to left is known as (trace), (retrace).
5. "Blanking" is necessary during (trace), (retrace).
6. In television a fundamental rate of _____ frames per second was chosen.
7. A television picture is composed of _____ lines.
8. There are two fields in one picture. True _____ False _____
9. Select the correct aspect ratio. (5 : 2), (3 : 4), (2 : 5), (4 : 3)
10. The audio signal is ________ modulated with the carrier. The video signal is ________ modulated with the carrier.
11. How many picture elements compose one picture?
12. What is the bandwidth of one channel?
13. Draw a block diagram of a monochrome transmitter and explain the functions of each block.
14. Draw a block diagram of a monochrome receiver and explain the functions of each block.

3 magnetism

The fundamentals of magnetism presented here are pertinent to the understanding of the chapters on sine-wave principles, picture-tube characteristics, and convergence.* By no means is it implied that this chapter presents all there is to know about magnetism. Nor should the student drop his study of the phenomenon, the heart of electronics.

Characteristics

The space surrounding a magnet, where magnetic forces act, is known as the magnetic field. A compass may be used to gain information about the magnetic field. Figure 3–1 shows the behavior of the needle of a compass used to explore the field surrounding a simple bar magnet.

* Chapter 3 adapted from *Fundamentals of Electronics,* Navpers-93400-A1B, 1965. Used with permission of the Bureau of Naval Personnel.

Notice that the compass needle aligns itself in various positions as it is placed at different points in the magnetic field. The alignment of the compass needle indicates a definite line of direction, or directional force, in the magnetic field.

The pattern of this directional force can be determined through a simple experiment with iron filings. A piece of glass is placed over a bar magnet and iron filings are sprinkled on the surface of the glass. The magnetizing force of the magnet passes through the glass, making each iron filing a temporary magnet. When the glass is tapped gently, the iron particles align themselves with the directional force in the magnetic field, just as the compass needle did previously. The filings form a definite pattern—a visible presentation of the forces comprising the magnetic field.

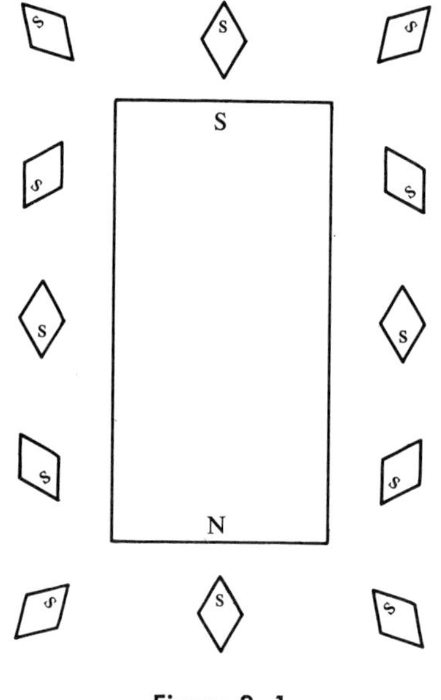

Figure 3–1

The arrangement of the iron filings in Figure 3–2 indicates that this magnetic field is very strong at the poles and weak away from them. It is also apparent that the magnetic field extends from one pole to the other, forming a loop about the magnet.

Lines of Force

To work easily with magnetism we use imaginary lines to represent the force existing in the area surrounding a magnet (Figure 3–3). These

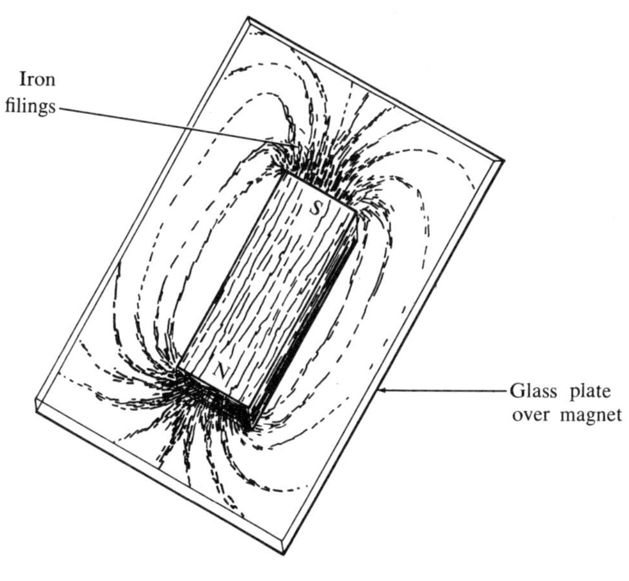

Figure 3–2

lines, called *magnetic lines of force,* symbolize the pattern of the directional force of the magnetic field. They emanate from the north pole of a magnet, pass through the surrounding space, and enter the south pole. The lines of force then travel "inside" the magnet from the south pole to the north pole, completing a closed loop.

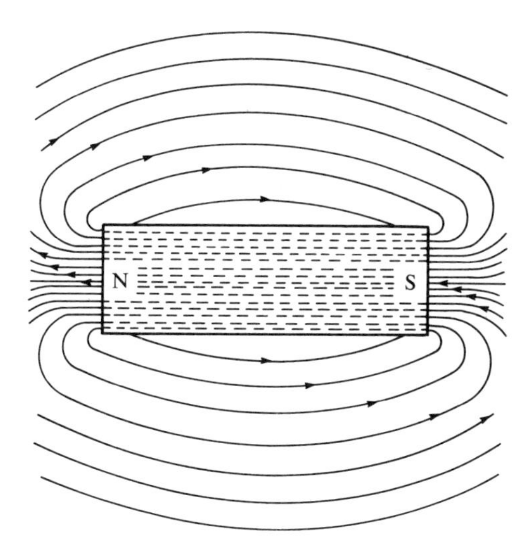

Figure 3–3

Although magnetic lines of force are imaginary, the understanding of magnetism can be simplified by assigning certain real properties to them. Lines of force can be compared to rubber bands which stretch outward when a force is exerted and contract when the force is removed. The properties of magnetic lines of force are as follows:

1. Magnetic lines of force are continuous and form closed loops.
2. Magnetic lines of force never cross one another.
3. Parallel magnetic lines of force, traveling in the same direction, repel each other. Parallel magnetic lines of force, traveling in opposite directions, unite, forming single lines which travel in a direction determined by the magnetic poles which create them.

The total number of magnetic lines of force leaving or entering the pole of a magnet is the magnetic *flux*. The number of flux lines per unit is *flux density*. The intensity of a magnetic field is directly related to its magnetic force.

Magnetic Shielding

There is no known insulator against magnetic flux. Any material will be penetrated by the passage of magnetic flux when placed within a magnetic field. For example, the glass placed over the bar magnet in the iron-filing experiment did not stop the penetration of the magnetic field—nor would have paper, copper, gold, or any other nonmagnetic material. Since it is not possible to block magnetic fields through insulation, needed protection from magnetic forces is obtained by redirecting the field. Because the sensitive mechanism of electric instruments becomes inaccurate when subjected to the influence of stray magnetic fields, such redirection is often necessary. Magnetic material such as soft iron may be used as a decoy in order to redirect the lines of force.. If soft iron is placed in a magnetic field, most of the lines of force take the easiest path, passing through the magnetic material and completing a closed loop (see Figure 3–4).

Because magnetic lines of force take the path of least opposition, when we surround an object with a material having a high permeability, the object is protected. A sensitive instrument is protected by enclosure in a soft iron case called a *magnetic screen,* or *shield,* as shown in Figure 3–5. It must be emphasized again that there is no insulator for magnetic

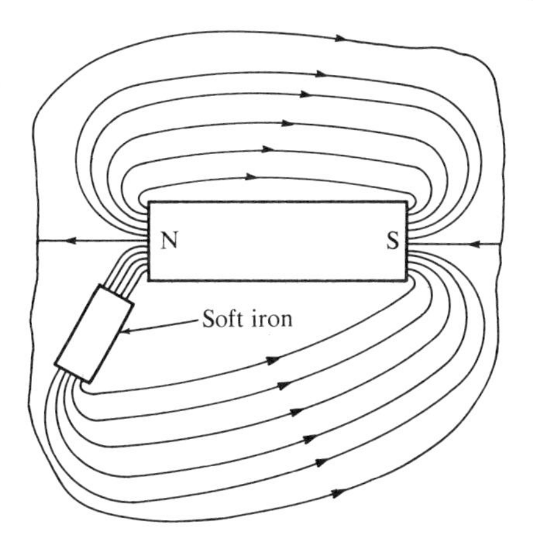

Figure 3–4

lines of force. Placement of an instrument inside an iron shield creates an insulating *effect.*

Electromagnetism

Scientists and physicists have performed many and varied experiments involving current and magnetism. Prior to 1820, it was thought that magnetic fields were unrelated to current flow. However, Hans Oersted (1777–1851), a Danish scientist and chemist, believed that there was a definite relationship between current and magnetism. He spent many

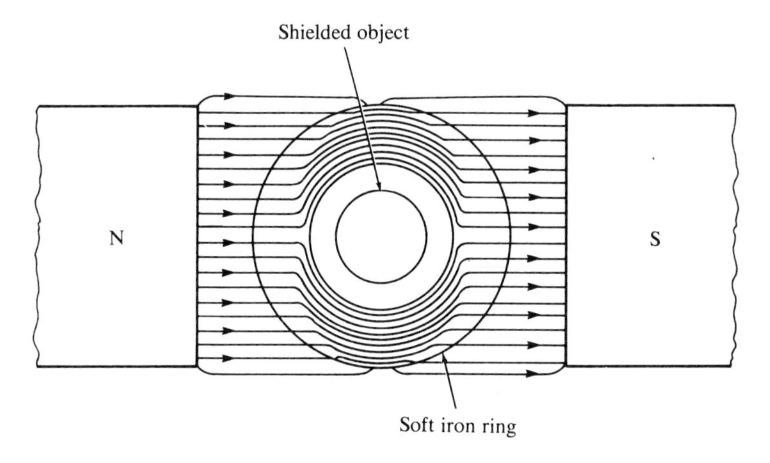

Figure 3–5

hours seeking this relationship. Unfortunately, Oersted and others labored under the misconception that the associated electromagnetic field would be exerted in the *same direction* as the current flow in a conductor. In trying to prove this idea, Oersted positioned a magnetic needle so that it lay at *right angles* to the wire. Oersted assumed the magnetic force would cause the needle to swing parallel to the wire. He was puzzled because no deflection of the needle occurred when current was forced through the wire. This was construed as proof that no relationship between magnetism and current existed.

Oersted came upon his discovery that the magnetic field exists at right angles to the current flow one day while teaching a group of students. He placed the magnetic needle parallel to the wire and noticed a deflection as current passed through the wire. Further investigation led him to formulate this theory of electromagnetism.

Left-hand Rule for Conductors

A simple rule, the left-hand rule, allows us to determine the direction of magnetic flux around a conductor when we know the direction of current flow. If the left hand is placed with the thumb pointing in the direction of electron (current) flow, the curled fingers will point in the direction of the flux lines encircling the conductor. The left-hand rule for conductors and the manner in which the intensity of the flux lines diminishes with distance from the conductor, are illustrated in Figure 3–6.

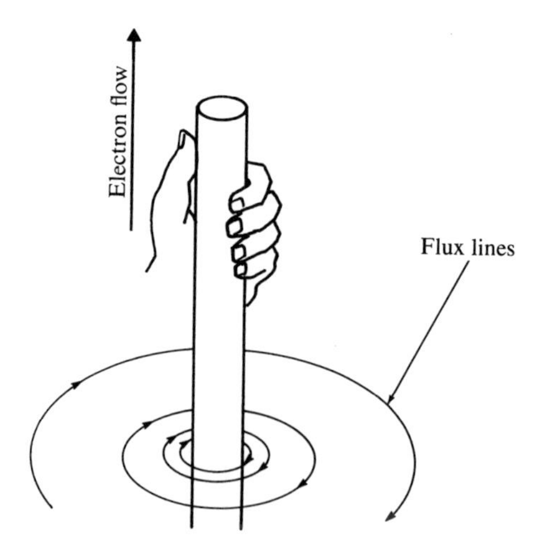

Figure 3–6

If the electron flow in Figure 3–6 is reversed and the left-hand rule is again applied, the thumb will point downward, indicating that the direction of the lines of flux is reversed. The left-hand rule may also be used to determine the direction of electron flow if the direction of the flux lines is known. Point the curled fingers in the direction of the circular flux lines and the thumb will point in the direction of electron flow.

Left-hand Rule for Coils

A coil is constructed of many individual current-carrying loops. These loops are placed close together causing their flux lines to interact as shown in Figure 3–7(b). The large number of flux lines is concentrated in the center of the coil. Figure 3–7(a) illustrates the interaction of flux lines in a coil by showing two segments of wire taken out of adjacent loops. Note first that the current travels the same direction in both conductors. Second, when the coils are not tight against each other, the flux lines very close to the wire still form closed loops around their own wire. Importantly, the outer flux lines interact and extend themselves to emcompass both conductors. As loops carrying current in the same direction are added, the magnetic lines of force merely extend themselves to provide a single field around the resulting coil. This is the internal action within a coil which allows lines of force to maintain closed loops around the entire coil.

Just as the left-hand rule indicates the relationship between current and flux in the single conductor and loop, there is a rule to indicate the relationship between current and flux in a coil. If the four fingers of the left hand are held (or wrapped) around the conductor of the coil pointing in the direction of electron flow, the thumb (held at right angles to the fingers) will indicate the direction of magnetic flux lines through the coil.

The end of the coil from which the flux leaves is called the north pole. Flux enters the coil at the south pole. The direction of current flow can be found, using the left-hand rule, when the poles of the coil are known. If the left hand is placed around the coil with the thumb pointing in the direction of the north pole, the fingers will point in the direction of current flow.

Figure 3–7(b) illustrates the magnetic field distribution around the coil and the left-hand rule for coils. Notice that the fingers point along the conductor in the direction of the electron flow. The thumb points in the direction of the flux lines through the center of the coil. Notice also that the flux lines leave the north pole and enter the south pole.

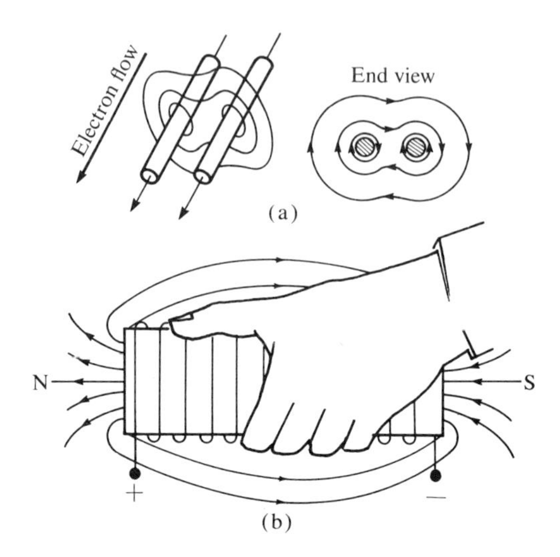

Figure 3–7

Left-hand Rule for Generators

Figure 3–8 illustrates the left-hand rule for generators. The vectors, or arrows representing direction and magnitude, show the mutually perpendicular relationship between three factors—electron flow, magnetic flux, and the conductor. (Notice the box in the left hand with the three vectors extending outward from the single corner in the palm. A quick glance at any square object or the corner of a book should help you to visualize the arrangement.) This vector arrangement is used whenever the left-hand rule for generators is applied.

In Figure 3–8, the thumb is pointing toward the top of the page along the vector marked "conductor." This indicates that the conductor is being moved from the bottom of the page toward the top, cutting the flux lines in the process. The forefinger is at right angles to the conductor and is pointing along the vector marked "magnetic flux." The direction of the forefinger indicates that the flux flows from north pole to south pole. The middle finger points to the bottom of the page along the vector marked "electron flow."

The direction of electron flow within a source is from positive to negative (north pole to south). Since even the movement of an open-circuited conductor through a flux field will cause a displacement of electrons (giving a deficiency of electrons at one end of the conductor and a surplus at the other), the middle finger points in the direction that electrons will move inside the loop. Since the section of conductor within the flux field will act as a source, a polarity is given to indicate the potential acquired from electron movement. If a circuit is connected as shown by the dotted line, its electron flow follows in the same direction, as indicated by the middle finger of the left-hand generator rule. Thus, when a complete circuit exists, the last line of the left-hand rule for generators is changed to read: *Then the middle finger will point in the direction of the electron flow.*

Figure 3–8 shows the effect of a change of direction in any of the factors involved. If the thumb is pointed down, but the flux lines are kept in the same direction, the conductor is indicated moving down through the field. The electron flow is now opposite its previous direction. If the thumb is pointed down, but the direction of the flux lines is reversed (by moving the forefinger around to point in the opposite direction), the direction of electron flow is again changed.

From this simple illustration we learn that change in the direction of the motion of either the conductor or the flux lines will reverse electron flow. Change in the direction of motion and flux lines will not change the direction of current flow.

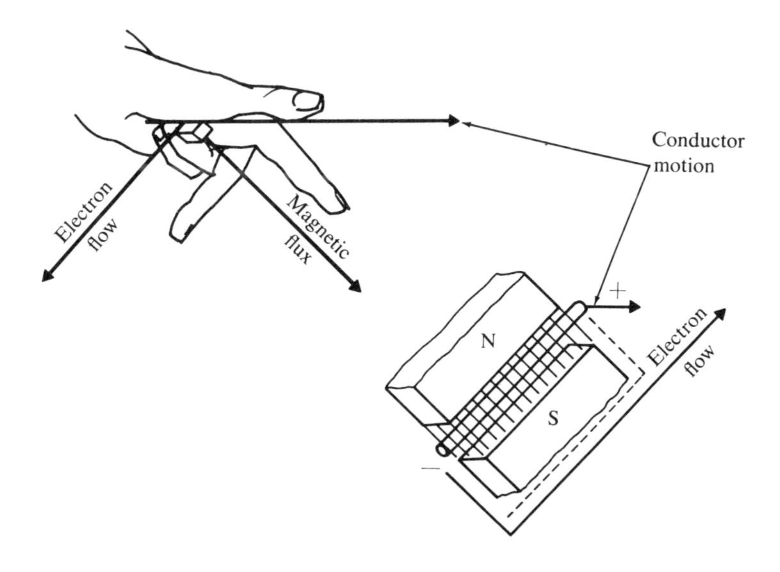

Figure 3–8

questions

1. Lines of force travel from the north pole of a magnet to the south pole. True _____ False _____
2. The total number of magnetic lines of force leaving or entering the pole of a magnet is called magnet __________.
3. The number of flux lines per unit is known as flux __________.
4. There are materials that will insulate against magnetic flux. True _____ False _____
5. Shielding of delicate instruments will block magnetic fields. True _____ False _____
6. Describe the left-hand rule for conductors.
7. Describe the left-hand rule for coils.
8. Describe the left-hand rule for generators.
9. List three characteristics of magnetic lines of force.
10. What is a magnetic field?
11. What did Oersted prove?

4 chromaticity diagram

Figure 4–7 shows a chromaticity diagram which is a chart of all the colors that can be seen by the human eye. The large horseshoe-shaped curve with the numbers plotted along its edge represents the colors appearing in a spectrum of sunlight, giving the wavelength of these colors in millimicrons. The diagram helps us find a color, as a road map helps us find a place.

Introduction to Vectors

The chromaticity diagram will best be understood with the aid of simple vectors. Vectors, as used here, do not have to be explained analytically by the application of trigonometry, but can be adequately explained graphically.

Suppose that we live in the center of the United States and our starting point is our home. Home is reference point zero. The map scale shows that one inch equals ten miles. We will, for simplicity, use the area designated by a vertical line due north from point zero and a horizontal line due east from the same reference point in Figure 4–1.

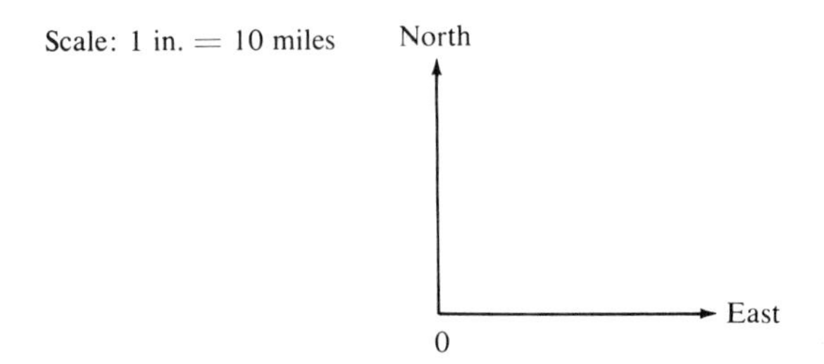

Figure 4–1

We get into our car and travel due north for ten miles. The trip is represented by a vector that has magnitude and direction, as in Figure 4–2.

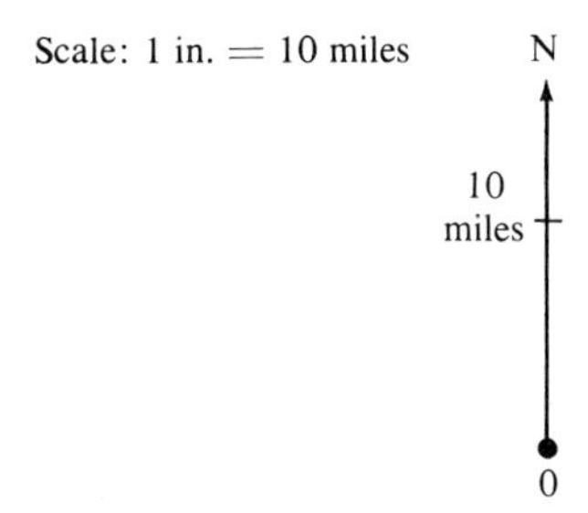

Figure 4–2

Now we turn our vehicle and travel ten miles due east. The horizontal vector also has a direction and magnitude, namely, ten miles due east (Figure 4–3).

Now we will find our distance and direction from home. To do this we draw a horizontal line from the ten-mile point of the vertical line, then a vertical line from the ten-mile point on the horizontal line (see Figure 4–4).

The point of intersection is our location from point zero or home. The actual distance from home is found by measuring the diagonal line connecting point zero to point A as shown in Figure 4–5. The line from point zero to point A measures 1.414 inches. This, multiplied by ten,

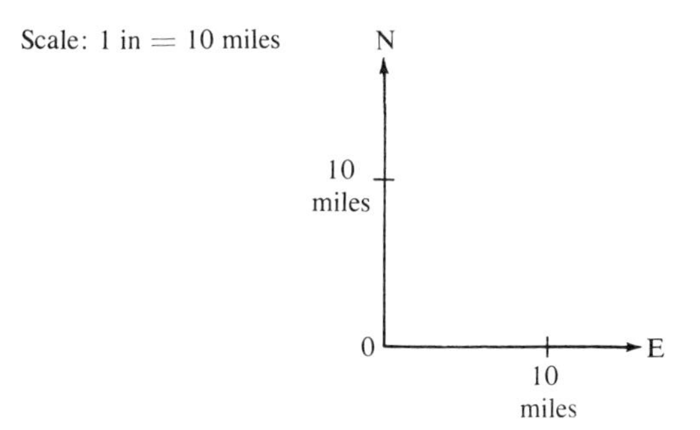

Figure 4–3

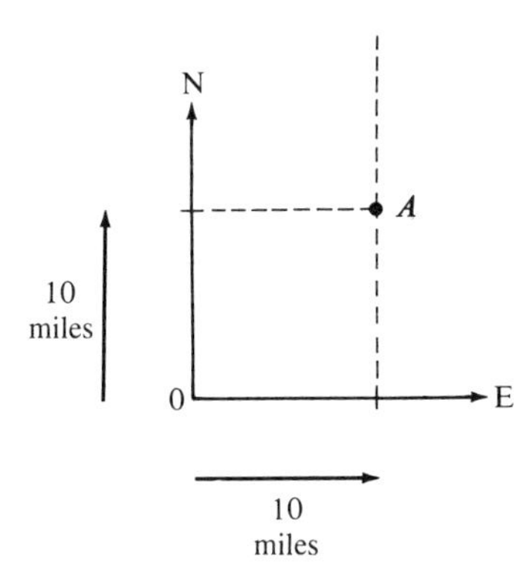

Figure 4–4

according to the map's scale of miles, gives us a distance of 14.14 miles northeast of home zero. With a protractor, we can measure the angle. To define our position with greater accuracy we can state that we are 14.14 miles from home zero and 45° north of east.

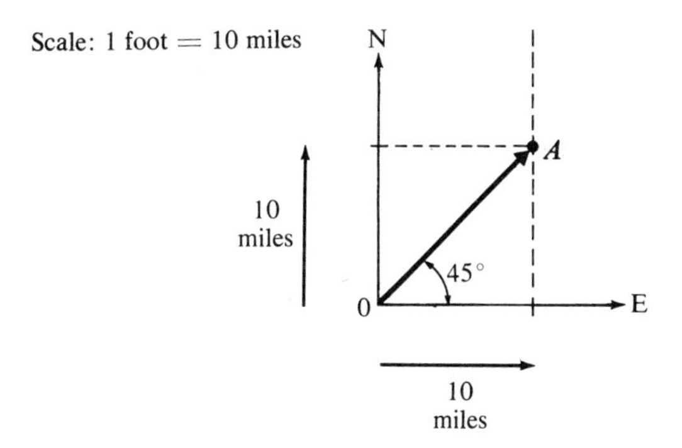

Figure 4–5

To summarize, the vector is used to locate our position from reference zero. The resultant vector, zero to **A,** was derived by adding two vectors of known magnitude and direction. The resultant vector **0** to **A** can logically be separated into the two vector components that created it (see Figure 4–6).

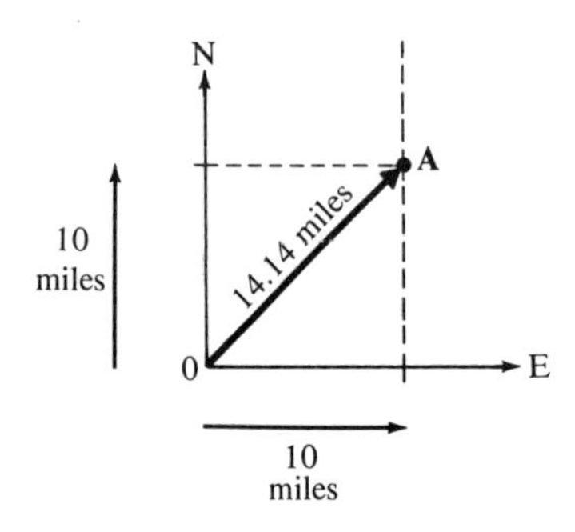

Figure 4–6

The Chromaticity Diagram

Let us utilize what we have just learned by going to a different type of "road map." The chromaticity diagram is the road map for color determination. Color, as previously stated, has three factors—hue, saturation, and brightness. To lay out a two dimensional representation of color, we will hold brightness constant. This is possible because brightness is independent of changes in hue and saturation.

Figure 4–7 shows the chromaticity diagram placed inside the x-axis and y-axis. Here, y takes the place of north and x replaces east. The chromaticity diagram takes the form of a horseshoe. By starting at the lower left corner of the horseshoe-shaped boundary and moving counterclockwise, we can trace a location for the full gamut of colors discussed in chapter 1. The outer boundary represents pure, 100% saturated colors starting at blue and progressing through cyan, green, yellow-green, yellow, orange, red, magenta, and back to blue. At the center of the horseshoe we see the area for white which represents 0% saturation. This area is identified by the coordinates $y = 0.316$ and $x = 0.310$. Between this point C and the outer periphery (100% saturated hues), the colors will range from low to high saturation.

The Color Triangle

The next step is to make *use* of the chromaticity diagram. Notice in Figure 4–8 that green, blue, and red form a triangle. Green is located

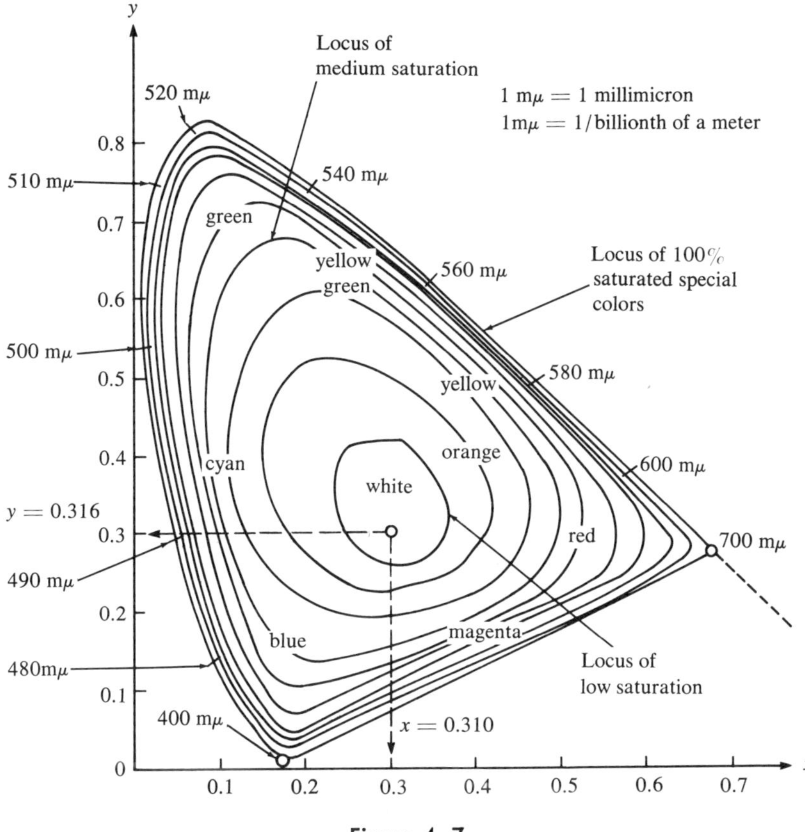

Figure 4–7

by the coordinates $x = 0.21$ and $y = 0.71$. Red is located by the coordinates $x = 0.67$ and $y = 0.33$. Blue is located by the coordinates $x = 0.14$ and $y = 0.08$.

The actual placement within Figure 4–8, does not encompass the complete area of the horseshoe. This results in a reduction of the possibility of reproducing all of the color hues that are available.

Green and red mixed together result in yellow (point *D*) in Figure 4–8. The more red mixed with the green, the more orange the color (halfway between points *D* and *C*). Mixing red and blue will result in magenta (point *F*). A green and blue mixture will result in cyan (point *E*). Combining yellow and blue will result in white (point *C*). Remember, yellow is a combination of red and green. The primary colors are red, blue, green.

The logic in choosing green as a primary color is evident in Figure 4–9. Flesh tones fall in the orange area between yellow and red. Green

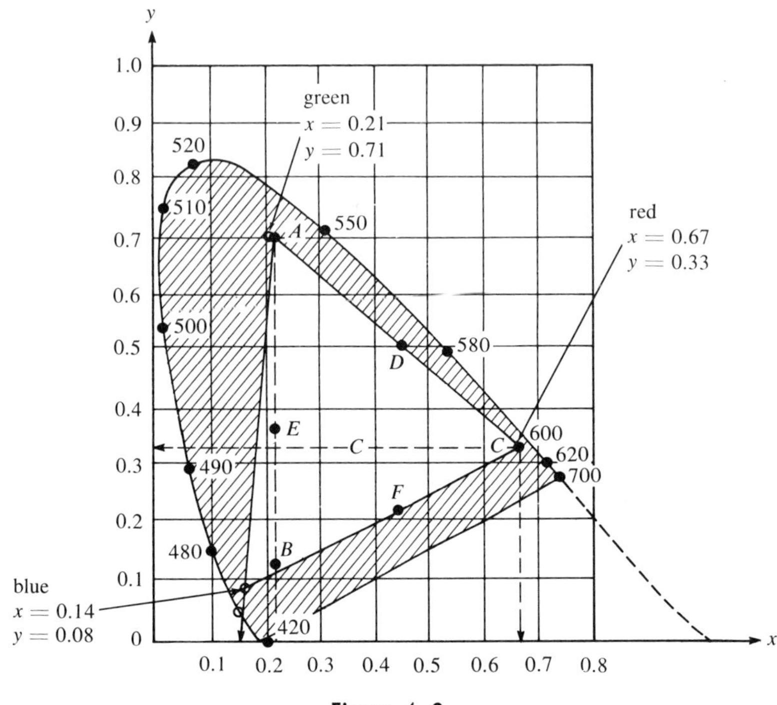

Figure 4–8

was chosen to keep the side of the triangle close to the outer boundary where saturation is at its maximum. If we move the triangle from green 1 to green 2, we would increase the triangular area but reduce saturation (purity of color) because the side of the triangle adjacent to yellow would move closer to center *C* or white. Previously it was stated that a 100%-saturated color contained no white light.

One might question why a fourth primary was not used to provide more color fidelity or exactness. The reason for this was strictly economic.

Finally, we have arrived at the point where it is possible to utilize the other two directions of the vector (Figure 4–10). Notice that the westerly direction is negative *x* and the southerly direction is negative *y*. It is now possible to locate our position in any direction from zero reference.

The color triangle in Figure 4–11 has been repositioned to put blue on the right. We can arbitrarily draw two intersecting axes on the triangle; one labeled the $B - Y$ axis, the other the $R - Y$ axis. Now we see that it is not always necessary to identify a color on the *x* and *y* plots of this chart because we can also identify any color in certain per-

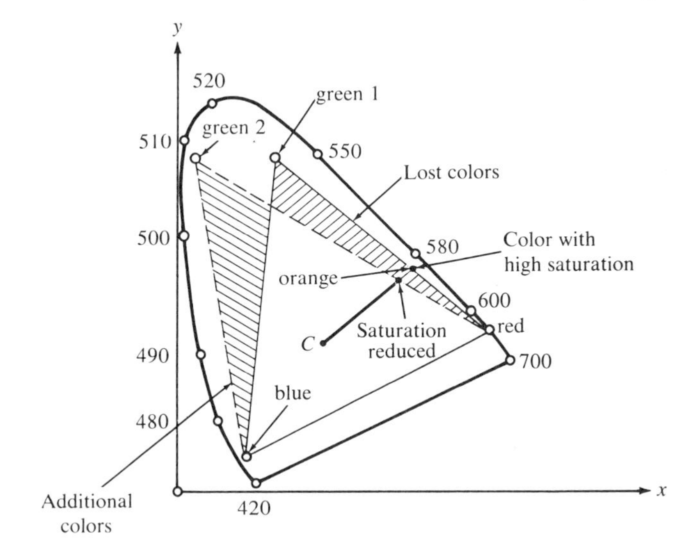

Figure 4–9

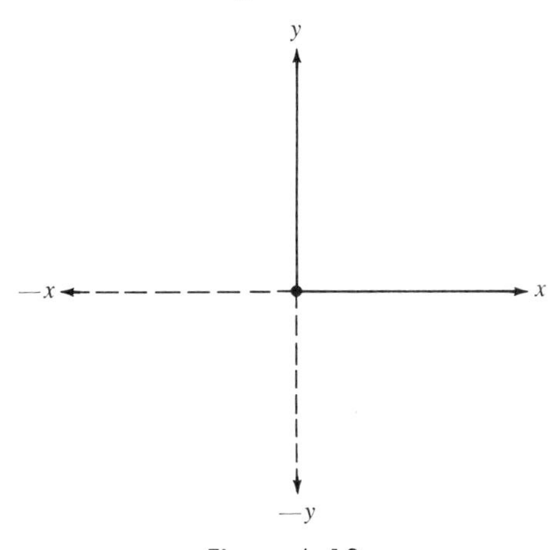

Figure 4–10

centages of the $B - Y$ axis and the $R - Y$ axis (Figure 4–11). A specific color red, for example, may be defined as $+0.7\ (R - Y)$ and $-0.3\ (B - Y)$. The terms $B - Y$ and $R - Y$ will be further defined in chapter 6.

Color Wheel

Let us cut a circle out of the center of the color triangle in Figure 4–8 and reshape it so that white is exactly in the center. Figure 4–12

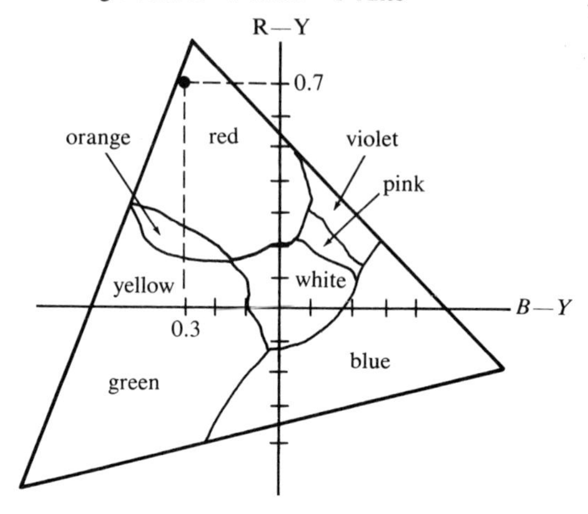

Figure 4–11

shows such a circle, referred to as a color wheel. It is also used to identify a specific color. Here we have the same relationships of color as we did on the chromaticity diagram—going from blue to cyan, to green to yellow, red, magenta, and back to blue. But instead of identifying the colors with x and y coordinates as we did on the chromaticity chart, we are identifying these colors in terms of the $R - Y$ axis. A magenta color can be identified as a certain percentage of the $R - Y$ direction and a certain percentage of the $B - Y$ direction.

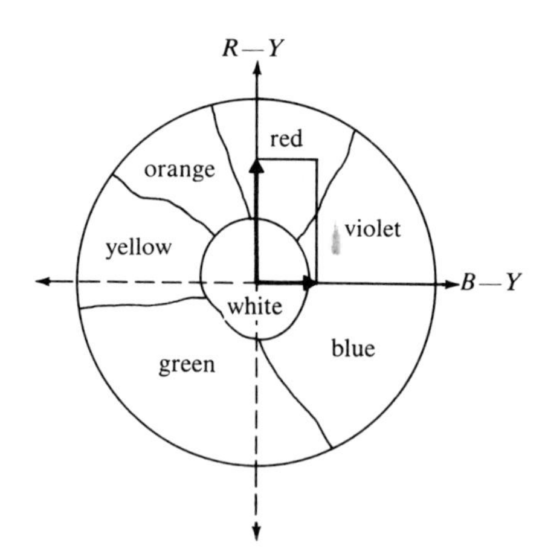

Figure 4–12

We can also identify a specific color of green in these terms. Figure 4–13 shows such a description. This green can be identified in units of a minus direction of the $B - Y$ axis and units of a minus direction of the $R - Y$ axis. With these percentages of the two axes we can, at any time, reconstruct or redefine that same specific color of green. With the

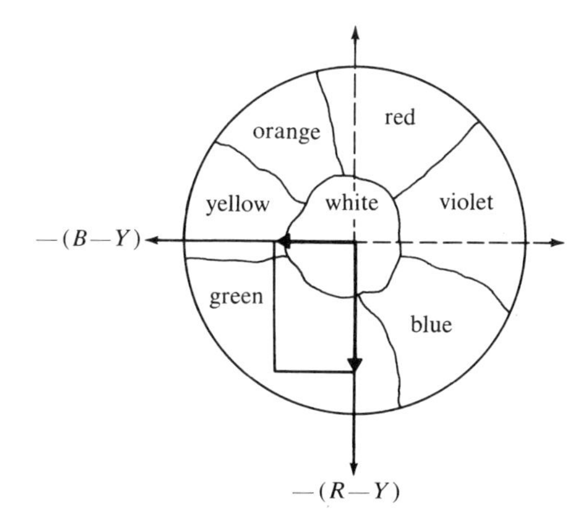

Figure 4–13

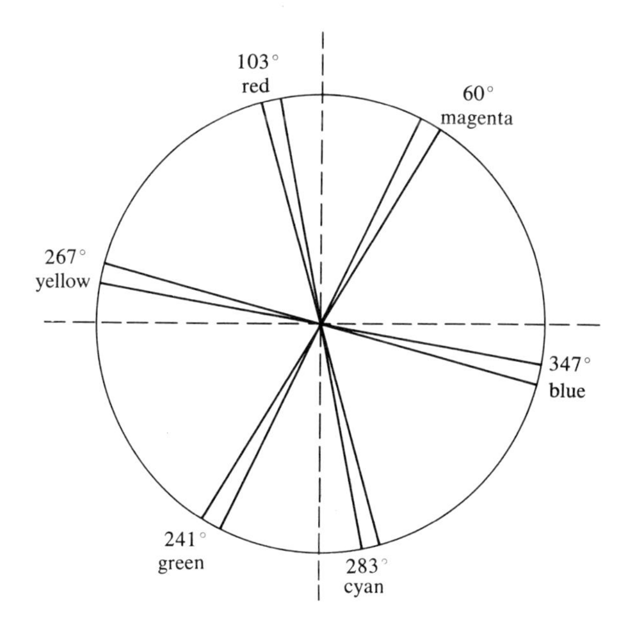

Figure 4–14

$R - Y$ and $B - Y$ axes we can define or locate any color on the color wheel simply in terms of minus or plus quantities.

Now that we have these colors represented along the circumference of a circle, we can also identify each color in terms of degrees of a circle. Figure 4–14 shows the color wheel with only the primary and complementary colors and the degrees of rotation where they appear.

Colors can now be identified in two ways: in degrees of rotation around the circumference of a circle, and in terms of plus and/or minus values of two axes at right angles to each other (quadrature). This arrangement makes it very convenient for us to use a vector rotating through 360° to identify a hue by its direction. The length of the vector indicates the amount of saturation.

questions

1. The road map for color determination is called the ________ diagram.
2. In Figure 4–15 locate the following:
   Red is at point _____
   Blue is at point _____
   Green is at point _____
3. Why is the color triangle kept close to the outer boundary of Figure 4–15?
4. A desaturated color would appear closest to what letter in Figure 4–15?

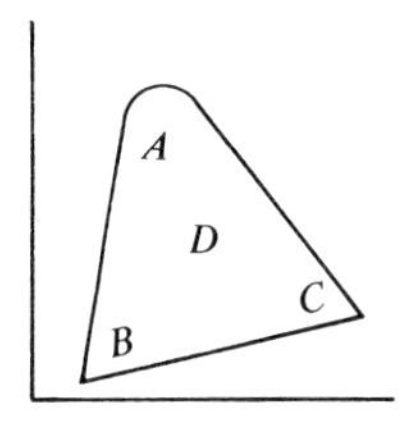

Figure 4–15

5. In Figure 4–7, give the coordinates to yellow, cyan, and magenta.
6. The outer boundary of Figure 4–15 describes what characteristic of light?
7. Flesh tones fall in the area between yellow and red.
   True _____ False _____
8. Pure 100% saturated color contains some white light.
   True _____ False_____
9. Using a fourth primary would provide additional color fidelity. Why was this feature omitted?
10. What are the two ways color can be identified?

5 sine-wave principles

Generation of a Sine Wave

Let us discuss briefly basic sine-wave principles utilizing our previous discussion of magnetism and some basic vector principles.

Figure 5–1 shows a loop of wire rotating counterclockwise in a magnetic field. Beneath each illustration is a vector representing the position (phase) of the loop in the magnetic field. As the loop rotates through the position in Figure 5–1(a), the black side of the coil moves toward the south pole, and the white side moves toward the north pole. Because the conductor is moving parallel to the direction of the magnetic field, no flux lines are cut and the induced voltage is zero. The meter which is connected across the output will read zero volts. The conductor is connected to slip rings which rotate with it. Stationary carbon brushes contact the slip rings and conduct the voltage to the meter.

When the conductor has rotated 90° from its initial position [Figure 5–1(b)], the black coil side moves downward and the white upward. Both sides cut a maximum number of flux lines and the induced voltage (indicated by the meter) is at a positive maximum. As the loop passes through the position shown in Figure 5–1(c), the coil sides again cut no flux lines, making the generated voltage zero. As the loop passes through the position shown in Figure 5–1(d), the coil sides cut a maximum number of flux lines making the generated voltage a negative maximum. The next 90° revolution of the loop completes the 360° revolution, making the generated voltage zero.

To summarize, as the loop makes one revolution of 360°, the induced voltage passes from zero to a positive maximum, to zero, to the negative maximum, and back to zero. If the speed of rotation is constant, the output voltage will be in a sine wave, as shown in Figure 5–2. If the time consumed for one revolution is one second, we say the frequency is one *hertz* (Hz), or one cycle per second. For example sixty hertz simply states that the frequency represents 60 revolutions per second.

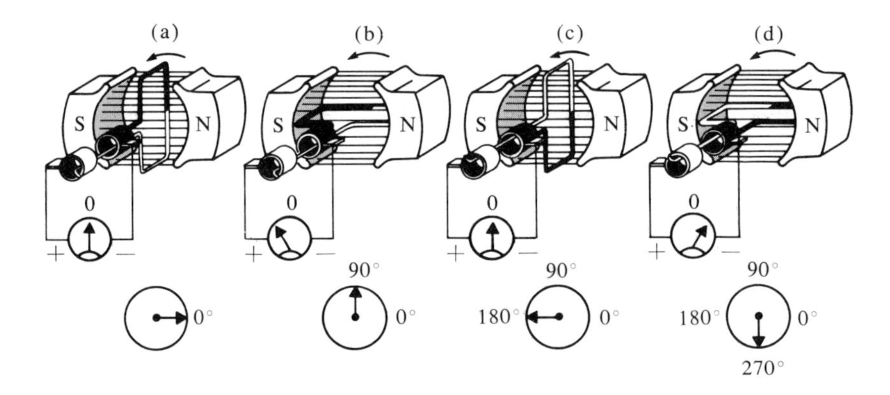

Figure 5–1

Vector Equivalent

The circle and sine-wave illustration (Figure 5–2) shows how a vector can be made to represent the phase and amplitude of a sine wave at any point in time. The vector pointing toward the zero-degree point on the circle corresponds to the zero-degree point on the graph of the sine wave. The vector proceeds counterclockwise to the 45° position, then on to the 90° position, and finally back to 0° position. The representative sine-wave phase and amplitude are indicated in the graph.

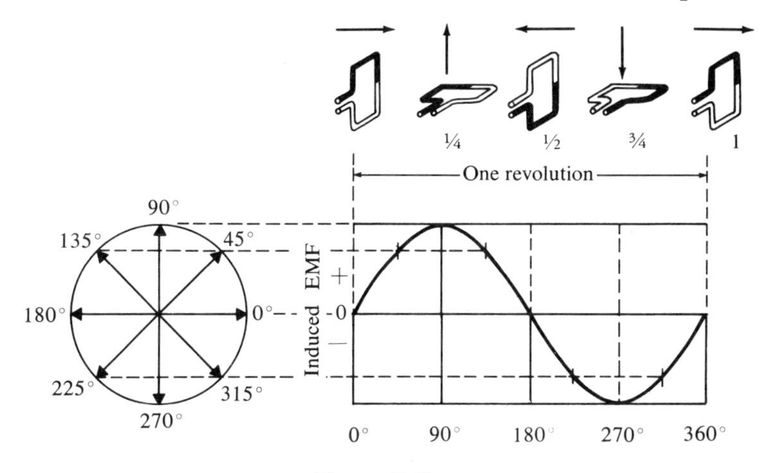

Figure 5–2

Let us now take two of these sine-wave generators, rotate them at exactly the same speed (frequency), but lead one ahead of the other by 90° in its rotation phase (position). This relationship is illustrated with vectors as in Figure 5–3. The armatures, or mechanical moving parts, that are producing these sine-wave voltages are turning at a speed that would make them difficult to observe, meaning that the vectors are turning at the same rate. In order to observe them and determine their action, we stop them, (figuratively), at the same instant on every cycle. We will use the vector as our standard and stop it at every cycle at the 0° position. This then describes the *A* sine wave shown in Figure 5–3.

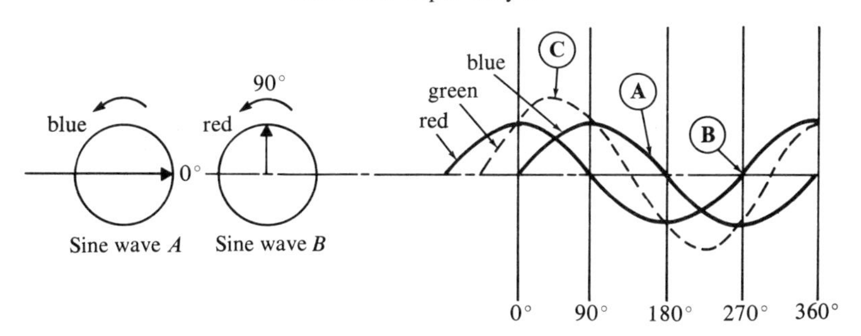

Figure 5–3

Vector **B** represents the generator that is leading (ahead of) vector **A** by 90°. This means vector **B** is 90° further along in its rotation in a counterclockwise direction than vector **A.** The output voltage described by this vector is shown by sine wave *B*. If we inject these two sine waves into an appropriate device, we can add them and produce a resultant sine wave that will have a phase different from either of the originals. This resultant is shown as the sine wave *C* in Figure 5–3.

A circuit that would give us an addition of the two original sine waves is shown in Figure 5–4 along with vectors indicating the two input phases and the resultant phase. The figure shows how we can explain the relationship between two or more sine waves by showing the direction (or phase angle) of their vectors.

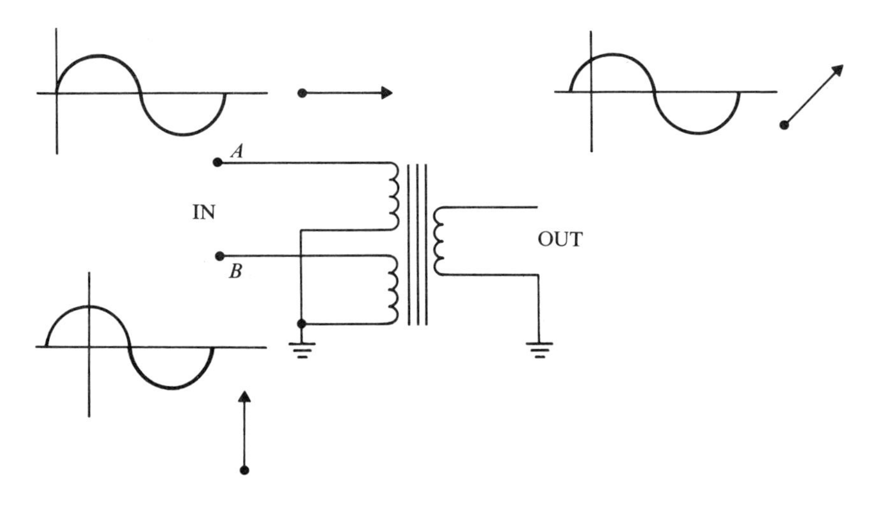

Figure 5–4

Combination of Vectors

Figure 5–5 shows how we combine vectors to get a resultant phase indication without the drawing of sine waves. In Figure 5–5(a), vectors **A** and **B** are two sides of a parallelogram. The diagonal becomes the resultant, or the phase angle, of the resultant sine wave.

The relative length of a vector is very important. While the direction of the vector indicates its phase, the length represents the magnitude or amount of sine-wave amplitude. If the length of vectors **A** and **B** [Figure 5–5(a)] each represent 1.0 volt, the length of vector **C** will equal 1.4 volts. Figure 5–5(b) shows how different-length vectors **A** and **B** change the phase and amplitude of resultant **C** within the limits of the parallelogram they form.

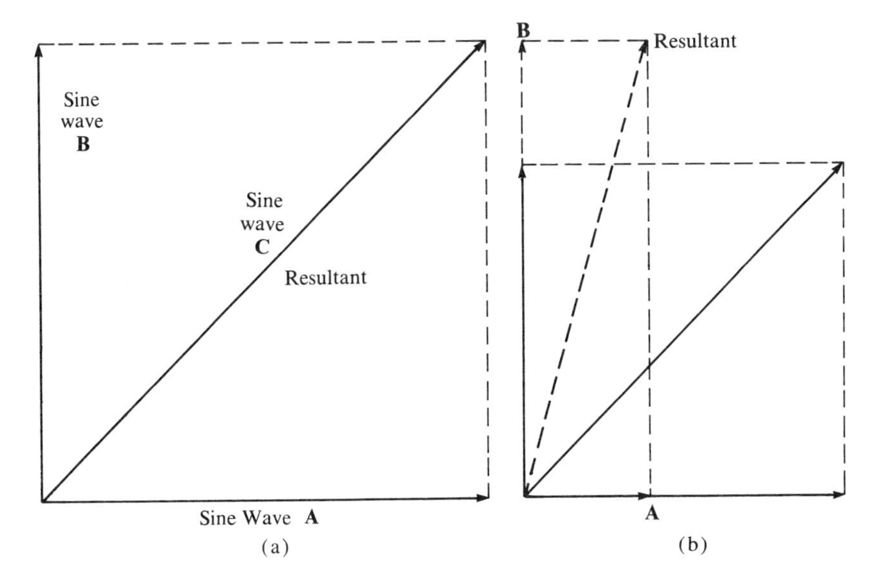

Figure 5–5

questions

1. The output of the generator shown in Figure 5–1 describes a ________ wave.
2. Illustrate how a vector can be made to represent the phase and amplitude of a sine wave at any point in time.
3. Draw the sine-wave equivalent of two vectors 90° apart.
4. Draw a circuit that would add two sine waves 90° apart and show its equivalent output.
5. Draw the vector addition of the resultant in Question 4.

6 standards

The sections of this chapter will build up bit-by-bit a compilation of the information that comprises National Television System Committee (NTSC) Standards. We will show how we build up the brightness signal and the color signal that will eventually describe the color scene that is transmitted. The brightness signal that results will achieve compatibility.

Brightness Signal

If we were to take an original image created by a group of red, green, blue, and white blocks and illuminate it with a white light, the blocks would reflect their respective colors as shown in Figure 6–1 (see also front endpapers).

Three separate black-and-white cameras are depicted in Figure 6–1, each with its respective color filter. The blue filter will pass only blue reflected light and will produce an output on camera A. The red filter will pass only the red reflected light and will produce an output on camera B. The green filter will allow the green reflected light to pass and will produce an output on camera C.

Each block is scanned by all three cameras at the same time but will reproduce only the light that is allowed to pass through the filters. When the red block is scanned, only camera B will produce an output, reproducing the red block on the color tube. When the green block is scanned, only camera C will produce an output, reproducing the green block on the color tube. When the blue block is scanned, camera A will produce an output and the blue block will be reproduced on the color tube.

Since the white object causes output from all three cameras when it is scanned, each of the guns in the color tube creates an image of the white object simultaneously. It is the electronic guns that create the images reproduced on the screen of the picture tube. In a later chapter we will describe the gun structure of a picture tube.

Due to the manner in which the eye responds to brightness (Figure 1–6), green appears 59% as bright as white. Red appears 30% as bright, blue 11% as bright. The color television reproduces this effect.

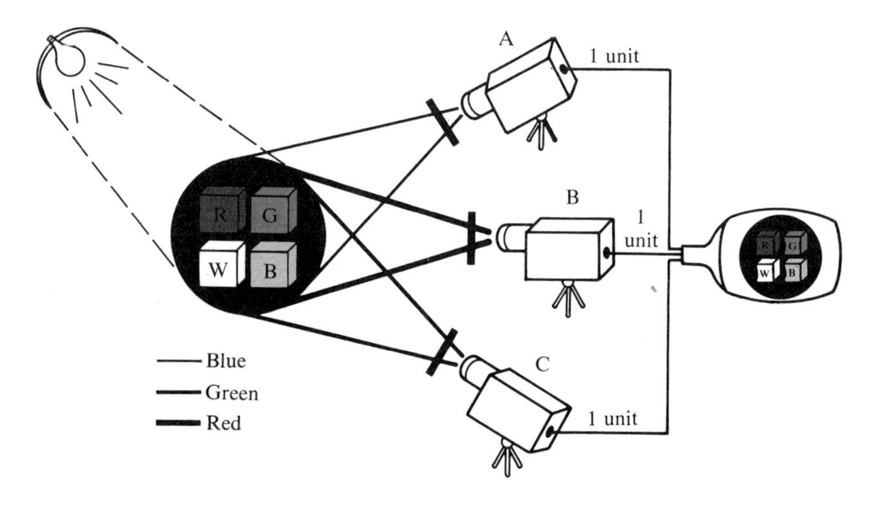

Figure 6–1

If the outputs of all three cameras were connected and applied to black-and-white television tubes instead of color tubes (Figure 6–2—see front endpapers), we would see the original scene, but in terms of brightness only. All the objects on tube A would appear to be equally white because the full camera outputs are applied through an adder circuit and fed to one gun of the black-and-white tube. We use the unit 1.0 to represent full output. The gun structure of any picture tube, whether color or black-and-white, reacts to the amplitude of the signal. One unit of white is equal to one unit of color, therefore, we see equal brightness levels. Compatibility problems would be created if a color signal were fed to a black-and-white tube with equal amplitudes. There would appear to be no difference between red and white. Red *should* appear as a shade of gray; white should appear as white. Therefore, the gun of tube B is not fed equal amplitude. Instead, it is fed 0.59 of green, 0.30 of red, and 0.11 of blue. These total a unit of 1.0. For white, all three proportions are applied to the gun. Now we have a reference of brightness. White is brightest. Green is 0.59 as bright as white, red is 0.30 as bright as white, and blue is 0.11 as bright as white. Notice tube B illustrates this difference in brightness.

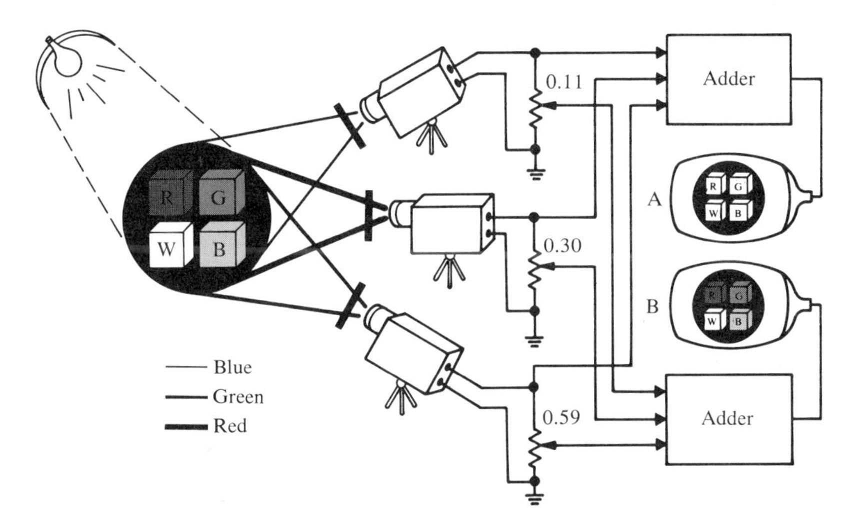

Figure 6–2

The brightness signal is called the *Y* signal. A formula that we will see frequently is $Ey = 0.30E_R + 0.59E_G + 0.11E_B$.

Simply stated, this formula says that the brightness voltage (Ey) is equal to 30% of the output voltage of the red camera, plus 59% of the output voltage of the green camera, plus 11% of the output voltage of the blue camera.

To summarize, the brightness signal (Ey) is transmitted in the manner used for black-and-white transmission, carrying high-detail information. A portion of the color signal has been used to produce this signal.

Addition of Color Information

Construction of the high-definition brightness signal is only the first step in a two-step process. The second is the addition of low-definition color information. This will result in the "splashing on" of low-detail color to a high-detail monochrome picture. A graphic approach will be used to illustrate this.

Figure 6–3 describes the output voltages of each camera for the given color bar. As stated previously, the voltage for red output (E_R) is described as a step from zero to +1.0 unit. The same holds true for green (E_G) and blue (E_B). When the color scene we are describing is white, all camera outputs are equal to one unit.

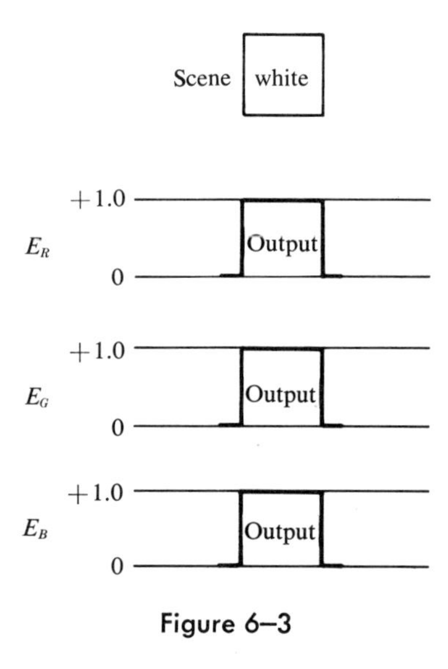

Figure 6–3

Red Output In Figure 6–4 red is the color bar that is being scanned. Only the red camera has an output of +1.0 unit. The green and blue have no outputs, equaling zero.

Green Output In Figure 6–5 green is the color bar that is being scanned. The green camera is the only one with an output (+1.0 unit), while the red and blue cameras have outputs both equal to zero.

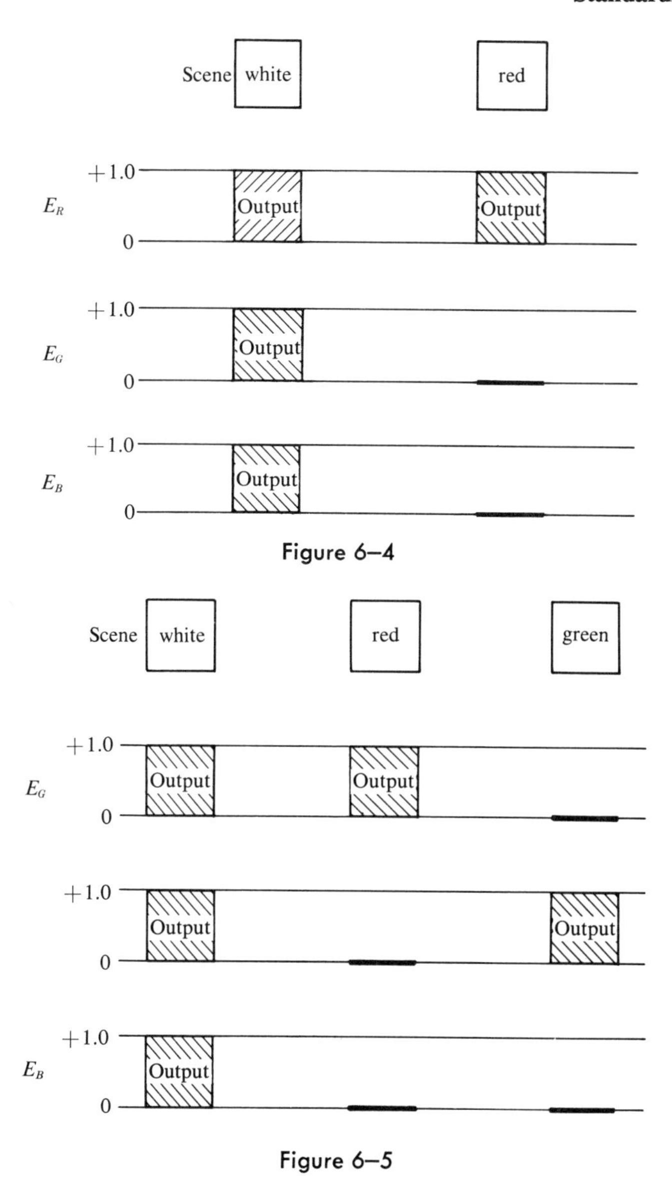

Figure 6–4

Figure 6–5

Blue Output In Figure 6–6 blue is the color bar that is being scanned. The blue camera is the only one with an output (+1.0 unit), while the red and green cameras have outputs equal to zero.

Yellow Output Figure 6–7 describes graphically the result of scanning a yellow bar. As was explained, yellow comprises a mixture of red

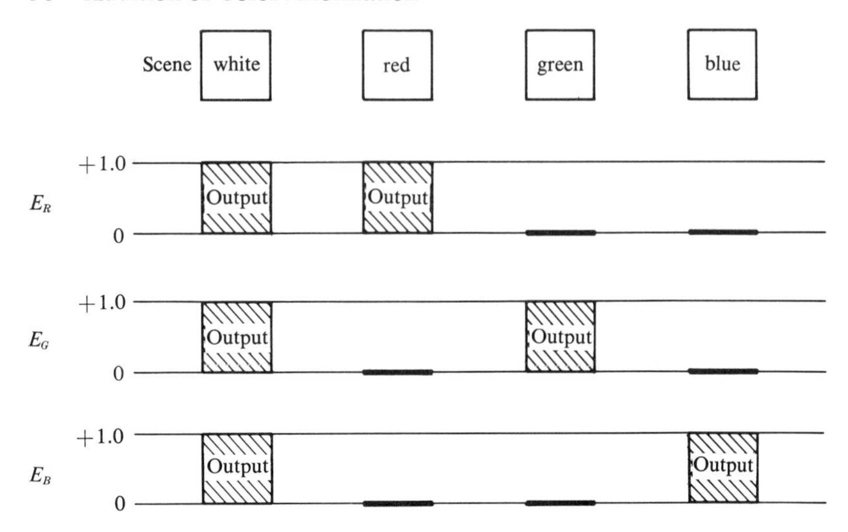

Figure 6–6

and green. Therefore, the output of the red camera is +1.0 unit and the output of the green camera is +1.0 unit. The blue output, as would be expected, is equal to zero.

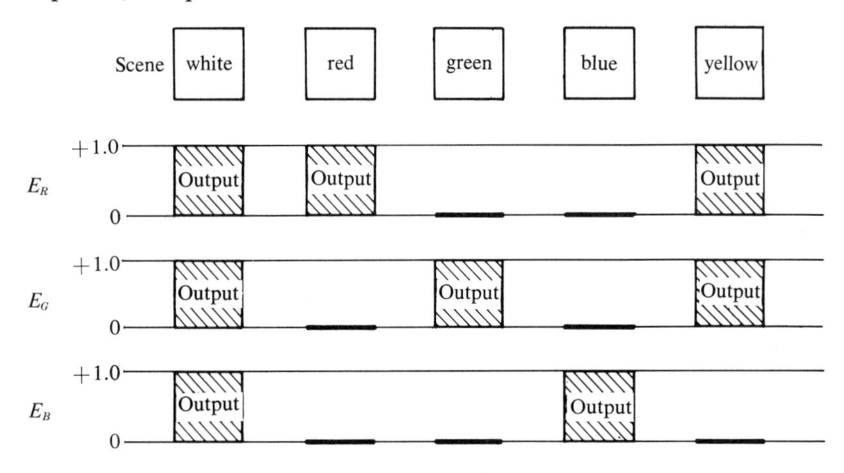

Figure 6–7

Brightness Signal Figure 6–8 shows the addition of the brightness signal *E*. The *Ey* signal for white provides one full unit. The graph shows that for *Ey* (running horizontally from left to right) the red camera provides 1.0 unit, only 0.30 of which is used. Next we see that the green camera provides 1.0 unit, but only 0.59 of that component is used. The blue camera also provides one unit, but only 0.11 of the blue output is used.

Now we have the overall components that satisfy the equation $Ey = 0.30R + 0.59G + 0.11B$. One unit is substituted for E_R, E_G, and E_B making the equation read $Ey = 0.30 \times 1.0 + 0.59 \times 1.0 + 0.11 \times 1.0$. This simplifies to $Ey = 0.30 + 0.59 + 0.11$. The solution for Ey in this equation is $Ey = 1.0$. Figure 6–8 shows the components that add to give us the requirement for the Y (brightness) signal for a white scene.

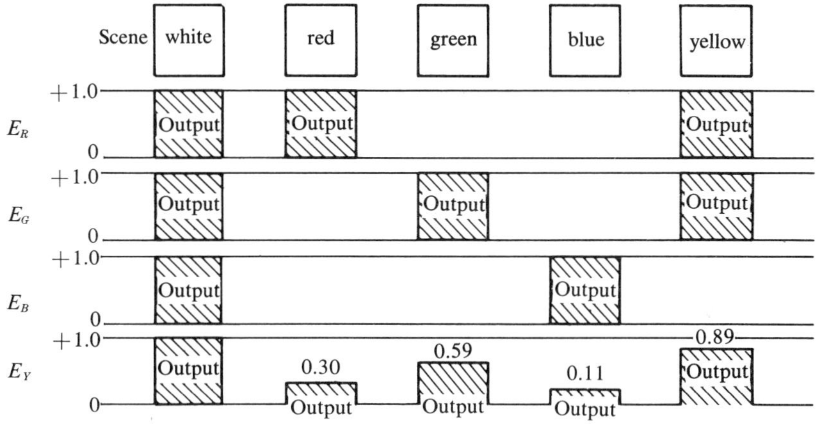

Figure 6–8

It is now possible to find what Y will equal when only a red signal is present. Starting from the red bar on top of Figure 6–8 and moving down vertically, we find the following:

$$E_R = 1.0 \text{ unit}$$
$$E_G = 0$$
$$E_B = 0$$
$$E_Y = 0.30$$

Another way of attaining the same result is to insert these quantities into the original formula for Ey.

$$Ey = 0.30E_R + 0.59E_G + 0.11E_B$$
$$Ey = 0.30 \times 1.0 + 0.59 \times 0 + 0.11 \times 0$$
$$Ey = 0.30$$

You should be able to do the same for green and blue. The solution for a green scene would be $Ey = 0.59$. The solution for a blue scene would be $Ey = 0.11$. Since Ey is the brightness signal, we have a reference of brightness for the primary colors as compared to white.

Before we get into the buildup of the color information, let us follow through the buildup of the signals required for a yellow scene.

$$\begin{aligned} \text{Yellow} = E_R &= 1.0 \\ E_G &= 1.0 \\ E_B &= 0 \\ Ey &= (0.30 \times 1.0) + (0.59 \times 1.0) + (0.11 \times 0) \\ &= 0.30 + 0.59 + 0 \\ Ey &= 0.89 \end{aligned}$$

Difference Signals

The chrominance, or color, signals are called the *difference signals.* Now we will see how these are developed. Figure 6–9 illustrates the difference-signal buildup for each color. By adding the color signal to the brightness signal, we find the color signal.

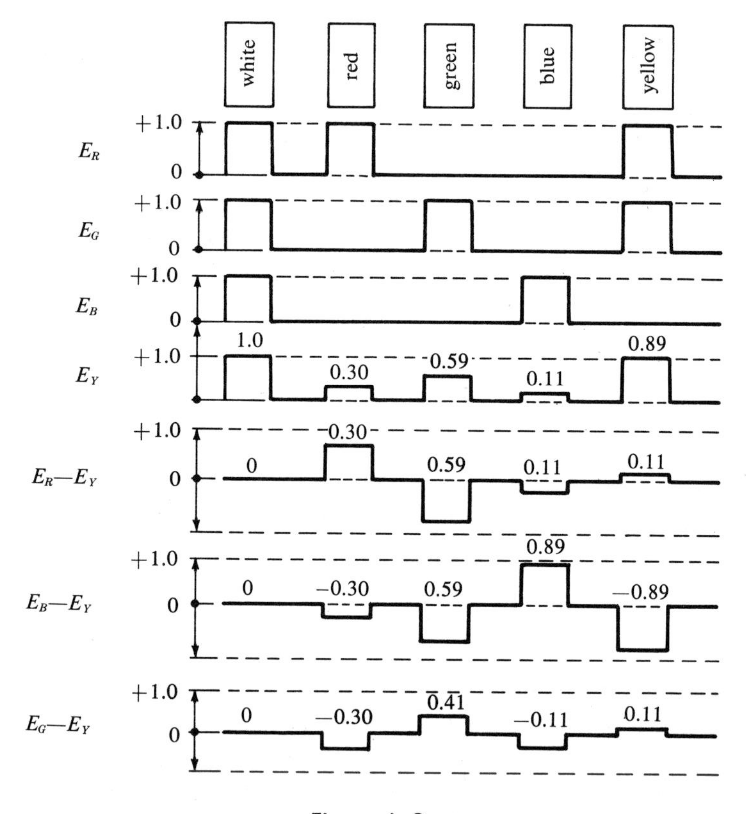

Figure 6–9

$$\text{Color Signal} + Ey = E_R$$

Therefore,

$$\text{Color Signal} = E_R - Ey$$

When this is done for the red, green, and blue outputs, the resulting color signals take the form shown in Figure 6–9: $E_R - Ey$, $E_B - Ey$, and $E_G - Ey$.

Figure 6–9 shows what these signals would be for each color. The steps for the white bar are as follows:

$$\begin{aligned} E_R &= 1.0 \\ E_G &= 1.0 \\ E_B &= 1.0 \\ E_Y &= 1.0 \\ E_R - Ey &= 1.0 - 1.0 = 0 \\ E_B - E_Y &= 1.0 - 1.0 = 0 \\ E_G - E_Y &= 1.0 - 1.0 = 0 \end{aligned}$$

The steps to follow for the *red* scene are:

$$\begin{aligned} E_R &= 1.0 \\ E_G &= 0 \\ E_B &= 0 \\ E_Y &= 0.30 \\ E_R - E_Y &= 1.0 - 0.30 = 0.70 \\ E_B - E_Y &= 0 - 0.30 = -0.30 \\ E_G - E_Y &= 0 - 0.30 = -0.30 \end{aligned}$$

The steps to follow for the *green* scene are:

$$\begin{aligned} E_R &= 0 \\ E_G &= 1.0 \\ E_B &= 0 \\ E_Y &= 0.59 \\ E_R - E_Y &= 0 - 0.59 = -0.59 \\ E_B - E_Y &= 0 - 0.59 = -0.59 \\ E_G - E_Y &= 1.0 - 0.59 = 0.41 \end{aligned}$$

The steps to follow for the *blue* scene are:

$$\begin{aligned} E_R &= 0 \\ E_G &= 0 \\ E_B &= 1.0 \\ E_Y &= 0.11 \\ E_R - E_Y &= 0 - 0.11 = -0.11 \\ E_B - E_Y &= 1.0 - 0.11 = 0.89 \\ E_G - E_Y &= 0 - 0.11 = -0.11 \end{aligned}$$

Notice that the difference signals are absent only for white scenes. For all other color scenes, the difference signals are present and are either positive or negative. Shortly, we will see how we use the brightness (Y) high-definition signal and the color low-definition signal.

So far we have discussed colors with 100% saturation only. Let us now develop the color signals when the saturation is less than 100%.

Figure 6–10 shows the wave form observed when three vertical bars—red, green, and blue—are scanned. In this case, however, there is some dilution of the colors. Here, the saturation is 50%. This means there is 0.5 unit of the other primaries present when a particular hue whose value is 1.0 unit is scanned.

Proceeding down the red bar vertically, as we did previously, leads to the following results:

$$E_R = 1.0$$
$$E_G = 0.50$$
$$E_B = 0.50$$

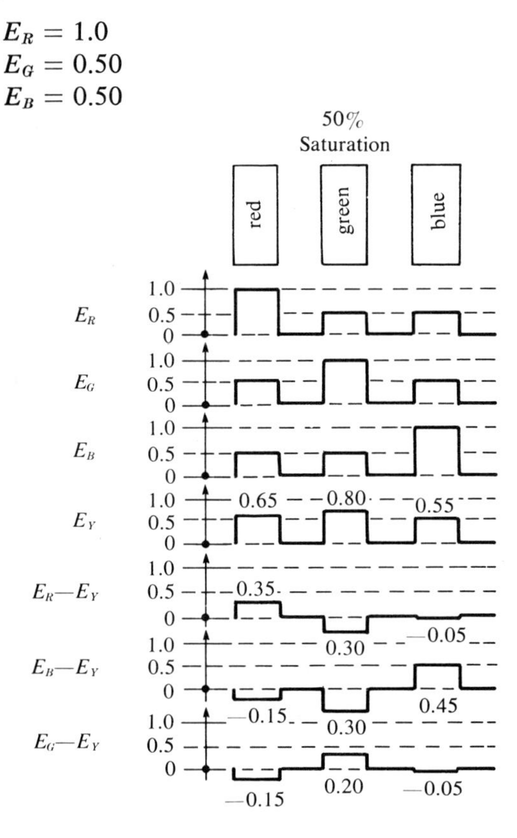

Figure 6–10

$$E_Y = 0.30E_R + 0.59E_G + 0.11E_B$$
$$= (0.30 \times 1.0) + (0.59 \times 0.5) + (0.11 \times 0.5)$$
$$Ey = 0.65 \text{ unit}$$
$$E_R - E_Y = 1.0 \quad - 0.65 = \quad 0.35$$
$$E_B - E_Y = 0.50 - 0.65 = \quad 0.15$$
$$E_G - E_Y = 0.5 \quad - 0.65 = -0.15$$

The same steps for the green and blue bar yield these results:

Green

$$E_R = 0.50$$
$$E_G = 1.0$$
$$E_B = 0.50$$
$$Ey = (0.30 \times 0.50) + (0.59 \times 1.0) + (0.11 \times 0.5)$$
$$Ey = 0.80$$
$$E_R - E_Y = 0.50 - 0.80 = -0.30$$
$$E_B - E_Y = 0.50 - 0.80 = -0.30$$
$$E_G - Ey = 1.0 \quad - 0.80 = \quad 0.20$$

Blue

$$E_R = 0.50$$
$$E_G = 0.50$$
$$E_B = 1.0$$
$$Ey = (0.30 \times 0.50) + (0.59 \times 0.50) + (0.11 \times 1.0)$$
$$Ey = 0.55 \text{ unit}$$
$$E_R - Ey = 0.50 - 0.55 = -0.05$$
$$E_B - Ey = 1.0 \quad - 0.55 + \quad 0.45$$
$$E_G - Ey = 0.50 - 0.55 = -0.05$$

It is apparent that the amplitudes of the color-difference signals are dependent upon the saturation of color being scanned. The exercise just shown can be performed to find the amplitude of the color-difference signals for any degree of saturation.

Color outputs are added by electronic circuits. Figure 6–11 shows the signals present in the transmitter during a red bar transmission. When all three cameras focus on a red scene, the green and blue cameras have no output. The red camera, however, has 1.0 unit of output which is fed directly to the adder. The adder is a circuit that combines different signals whose resultant is the addition of the individual signals. Through the voltage divider, 30% is fed to the brightness (Y) adder. Notice the blue output is fed to the blue adder. The green and blue voltages, divided 59% for green and 11% for blue, are also fed to the brightness-signal adder. The output of the brightness adder develops the Ey signal which, as stated before, is our high-definition signal. In this case, Y is equal to 0.30. The brightness-signal adder has another output which is fed to an inverter that converts the $+Y$ signal to a $-Y$ signal. The $-Y$ signal is fed to the blue adder and the red adder. Since the blue (E_B) signal is zero, its addition to the $-Y$ signal results in the equation $B - Y = -0.30$. The red signal, with its output of 1.0 unit, is added to the $-Y$ signal which yields $R - Y = 0.70$.

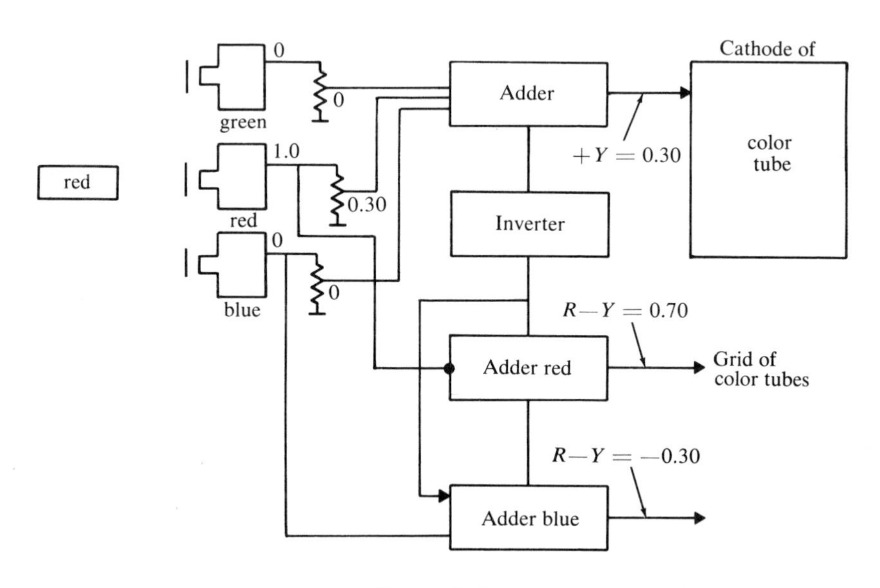

Figure 6–11

Elimination of $E_G - Ey$

Note the absence of the $E_G - Ey$ difference signal in Figure 6–11. It is possible and necessary to eliminate transmission of the $E_G - Ey$ difference signal for the sake of compatibility. Present-day monochrome television makes use of an effective pictorial bandwidth of approximately 4 MHz (as shown in Figure 6–12), making information available at the rate of approximately 8,000,000 elements of picture content. Color television systems must offer equivalent performance. Since our color system consists of three primaries it would be expected to have a 4 MHz pictorial bandwidth for each. This would require a TV channel to provide 12 MHz of bandwidth. Since monochrome channels have only a 4-MHz bandwidth available, and compatability is necessary, a 12-MHz bandwidth for transmission of color information is definitely out of the question. Since low definition of the color signal has been found adequate for public acceptance, it is advisable to drop one of the color-difference signals, the $E_G - Ey$ signal.

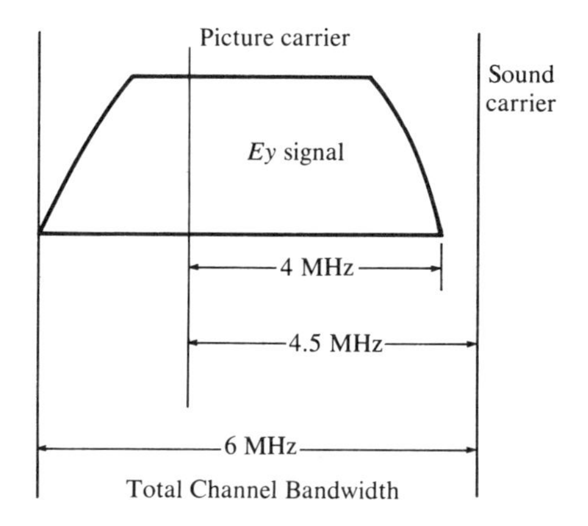

Figure 6–12

The means for omitting the $E_G - Ey$ signal is explained in Figure 6–13, which shows the three difference signals for red, green, and blue. By reversing the phase of $E_R - Ey$ and removing 51%, −0.51

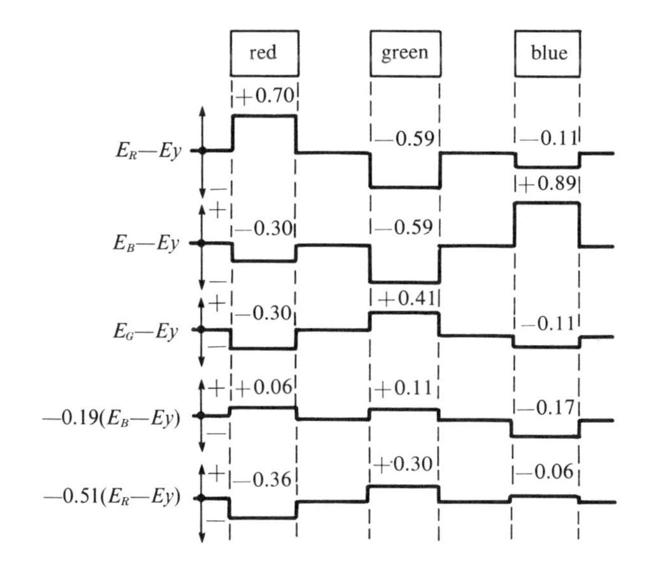

Figure 6–13

$(E_R - Ey)$ is obtained. We obtain $-0.19\ (E_B - Ey)$ in the same manner. Notice, if the two wave forms are added, $E_G - Ey$ results. Let us look at the red bar as an example.

$$\begin{aligned} E_B - E_Y &= -0.30 \\ -0.19\ (-0.30) &= +0.06 \\ E_R - E_Y &= +0.70 \\ -0.51\ (+0.70) &= -0.36 \\ +0.06 + (-0.36) &= -0.30 \\ E_G - E_Y &= -0.30 \end{aligned}$$

This exercise can be repeated for green and blue to prove that addition of the proportions stated yields $E_G - Ey$. The red bar in Figure 6–13 corroborates these quantities.

Difference signal $E_G - E_Y$ is developed in the receiver (Figure 6–14). The picture tube then acts as an adder. Remember, we must reproduce a red color with a unit of 1.0 to duplicate the transmission. In the following steps we will add Y to all the difference signals (Figure 6–14):

$$\begin{aligned} Y + (R - Y) &= 0.30 + 0.70 \\ &= 1.0 \text{ unit} \\ Y + (G - Y) &= 0.30 + -0.30 \\ &= 0 \\ Y + (B - Y) &= 0.30 + -0.30 \\ &= 0 \end{aligned}$$

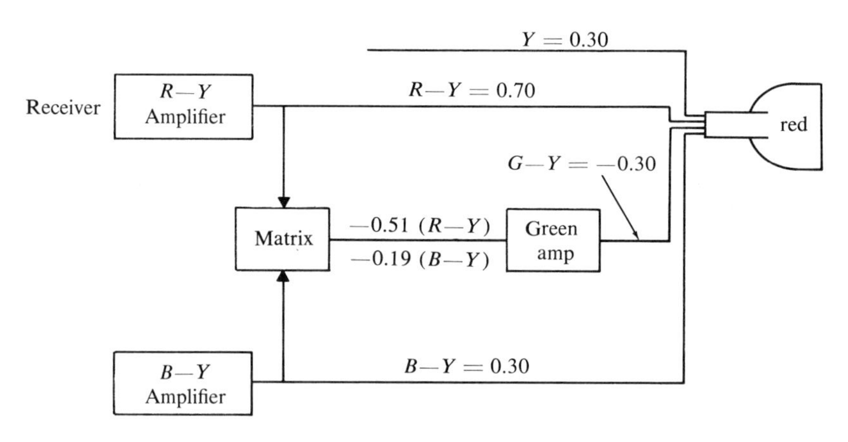

Figure 6–14

This addition at the picture tube indicates that the only color present is red, since it has a quantity of 1.0 (indicating 100% saturation). The green and blue components are equal to zero, indicating there is no color information present for these colors.

Thus, one transmission problem within the 4-MHz bandwidth has been eliminated by the omission of one color-difference signal.

Interleaving

As shown in Figure 6–12, the 6-MHz band is occupied by the brightness and the audio signals. Two color-difference signals ($B - Y$ and $R - Y$) which must be placed somewhere within this 6-MHz band, still remain. The existence of voids in the video spectrum makes placement of the two signals possible.

All the information in the video spectrum is carried by energy at separated frequency intervals. The remainder of the spectrum is empty and unused. The separated frequency intervals have a definite relationship to the synchronizing frequencies used in the television signal. Figure 6–15 illustrates a magnified monochrome signal. The predominant frequency interval is 15,750 Hz per second, which is the horizontal-scanning frequency. Surrounding these energy points are smaller amounts of energy separated by 60 Hz (vertical-scanning frequency).

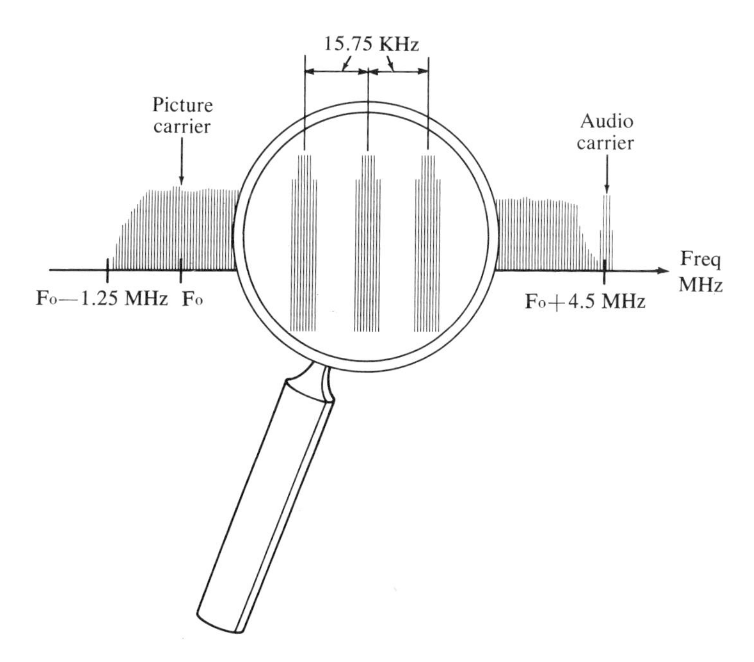

Figure 6–15

$B - $ Y and $R - Y$ information is inserted into the voids through selection of a subcarrier frequency of a multiple of half the line frequency. Figure 6–16 shows empty brightness (Y) energy points available between the picture carrier and the sound carrier.

Figure 6–17 shows the chrominance information interleaved, or inserted, in the empty spaces. These frequencies were chosen because they

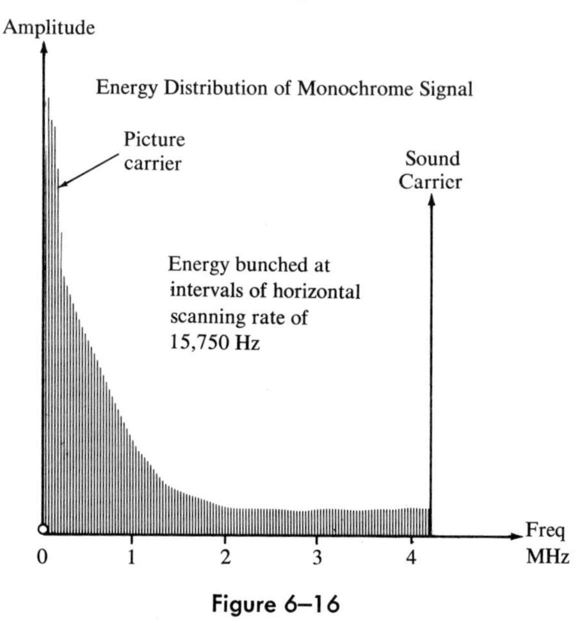

Figure 6–16

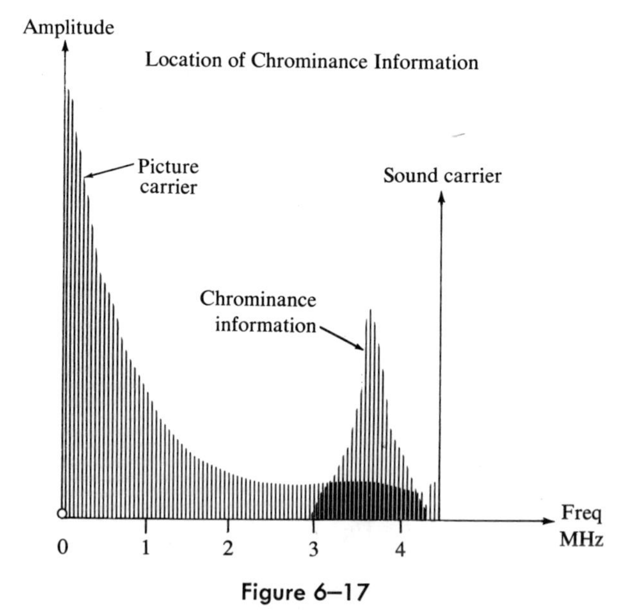

Figure 6–17

do not create interference between the monochrome and the chrominance information. The horizontal line frequency is 15,734.26 Hz. (The monochrome standard for line frequency is 15,750 Hz.) The new field frequency of 59.94 Hz maintains the standard 525 lines and uses the two-to-one interlace characteristics, remaining within the tolerance of monochrome standards. The color subcarrier was chosen to be 3.579 MHz to successfully obtain interleaving.

Figure 6–18 shows how the color synchronizing signal is transmitted. On the back porch of each horizontal sync pulse is placed a minimum of 8 cycles of reference 3.58-MHz signal. In chapter 7 we will see that the color sync signal is generated by the same oscillator that generates the exciting voltages for the balance modulators.

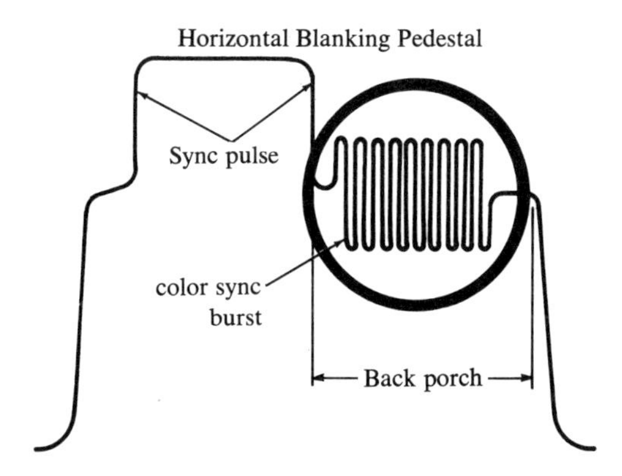

Figure 6–18

questions

1. The output of each camera for white is equal in amplitude.
   True _____ False _____
2. The brightness signal is identified by the letter _____.
3. The formula for Ey is equal to __________.
4. Ey for a white signal is equal to: (1.0 unit), (0.5 unit), (0 unit), (3 units).
5. Ey for red is equal to __________.
6. Ey for blue is equal to __________.
7. Ey for green is equal to __________.
8. Difference signal $E_R - Ey$ for white is _____%.
9. Difference signal $E_B - Ey$ for red (100% saturation) is _____%.
10. Difference signal $E_G - Ey$ for green (100% saturated) field is _____%.
11. Ey for a white field is equal to:
    ________ green + ________ red + ________ blue.
12. Which of the following difference signals do we use to modulate the transmitted signal carrier? $(E_R - Ey)$, $(E_B - Ey)$, $(E_G - Ey)$
13. By adding $-0.19\ (E_B - Ey) + (-0.51)\ (E_R - Ey)$ we get ______ difference signal.
14. The horizontal-sweep frequency for color is _____.
15. The vertical-sweep frequency for color is _____.
16. The subcarrier frequency is _____.
17. What is interleaving?

7 modulation

Types of Modulation

The process of superimposing information on a carrier wave is known as *modulation*. The carrier relays information from the transmitter to the receiver. There are several types of modulation. For instance, if the amplitude of the carrier is varied, but its frequency remains the same, it is said to be *amplitude-modulated*. If the amplitude of the carrier is constant, but the carrier frequency is varied, it is said to be *frequency-modulated*. When frequency and amplitude are constant, but the phase is changed, the carrier is said to be *phase-modulated*.

The color receiver utilizes all three methods of modulation. The audio system of a television receiver uses FM modulation. This holds true for black-and-white and color television receivers. The video carrier is amplitude-modulated and the color modulators are both amplitude-modulated and phase-modulated.

Figure 7–1 illustrates the basics of frequency modulation. It should be noted that the carrier wave is constant in amplitude and frequency [Figure 7–1 (a)]. This is known as the center frequency. The modulating audio frequency [Figure 7–1 (b)] will be used to modulate the carrier by varying the carrier frequency. The result of this action is shown in Figure 7–1 (c). Figures 7–1 (d) and 7–1 (e) illustrate the result of FM modulating the carrier by a higher audio frequency. When a modulating signal is applied, the frequency change above or below the center fre-

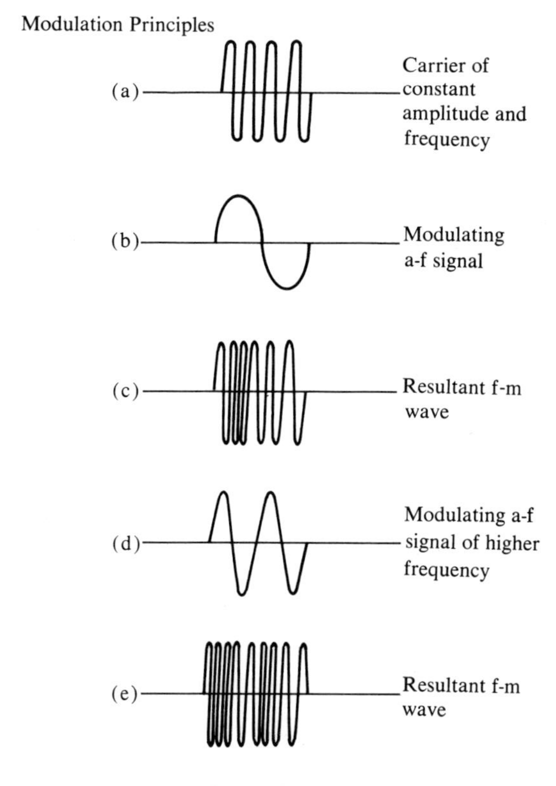

Figure 7–1

From Elements of Electronics *by Hickey & Villines.*
Copyright 1970 by McGraw-Hill Book Company.
Used with permission of McGraw-Hill Book Company.

quency is known as *frequency deviation.* The total variation between the minimum and maximum values of frequency is known as *carrier swing.*

Suppressed Carrier Modulation

At this point let us inject one more term—*suppressed-carrier modulation.* Later we will see that a common carrier (3.58 MHz) is amplitude-modulated by two different signals and separated in phase by 90°. The carriers are suppressed (canceled out) and the resultant modulated (AM) signal is sent to the transmitter. At the receiver this carrier is re-inserted to obtain the desired recovery of color information.

Amplitude Modulation

A clear understanding of amplitude modulation is extremely important in the understanding of color television. To repeat, modulation in its simplest form is the superimposing of information onto a carrier at frequencies between 60 Hz and 4 MHz. These are lower than the carrier frequencies. When a carrier is modulated by an audio tone for example, additional signals are produced in the output of the transmitter. These new frequencies are called *side-band signals.* One signal would be equal to the sum of the carrier and the modulating frequency. Another would equal the difference between the carrier and the modulating frequency. For example, if the carrier is 800 kHz and the modulation frequency is 2 kHz, the two side-band signals are 802 kHz and 798 kHz.

Vector Equivalent

Figure 7–2 shows a vectorial presentation of the 800-kHz carrier and the side bands produced by modulation. The carrier is represented by vector **OA;** vectors **AB** and **AC** represent the lower and upper side bands. We assume that each side band is half the length of the carrier. This is considered 100% modulation. The entire vector diagram is rotating around reference point **O** at the carrier frequencies. Vector **OA** can be considered to be standing still because its amplitude remains constant and because we are considering the effect of the modulation vectors on the amplitude of the carrier vector. The upper side band, vector **AC,** will rotate around point *A* in the counterclockwise direction since its frequency is higher (faster) than the carrier frequency. The lower side band, vector **AB,** will rotate around point *A* in the clockwise direction since its frequency is lower than the carrier and is turning slower.

An analogy may hely you visualize this. Consider three trains traveling side by side on parallel tracks. One train is traveling at 49 miles per hour. The center train is traveling at 50 miles per hour. The other train is traveling at 51 miles per hour. To the passenger on the center train, the first train will appear to be traveling backward at one mile per hour. The other train will appear to be traveling forward at one mile per hour. The passenger's train will feel as if it is standing still.

Now return to the rotating side-band vectors of Figure 7–2. Here, the resultant of the side bands adds vectorially to the carrier at every instant, producing the resultant modulated wave.

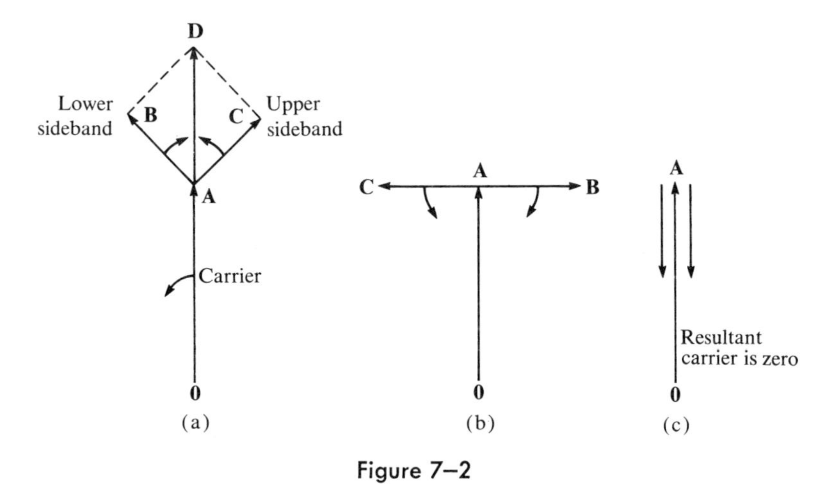

Figure 7–2

Effect of Sine-Wave Modulation on Carrier

Figure 7–3 considers the effect of a full cycle of sine-wave modulation on the carrier.

Position (1). The side bands cancel and the carrier is unchanged.

Position (2). The resultant of the side bands adds to the carrier, producing some intermediate point of the modulated envelope (modulation frequency information).

Position (3). The side bands, in phase, add to produce maximum envelope amplitude (or maximum height of the modulating signal).

Position (4). The side bands again add to the envelope amplitude as in position (2), but due to the rotation of the side-band vectors, they have exchanged positions.

Position (5). The side bands again cancel so that the carrier is not changed.

Position (6) The side bands subtract from the carrier to produce some intermediate point of the modulated envelope.

Position (7). The side bands are again in phase, but this time their resultant is 180° out of phase with the carrier, producing the trough, or complete cancellation, of the carrier necessary for 100% modulation.

Position (8). The side bands again subtract from the carrier as in position (6).

Position (9). The side bands have completed one cycle of rotation and are in the same phase as in position (1).

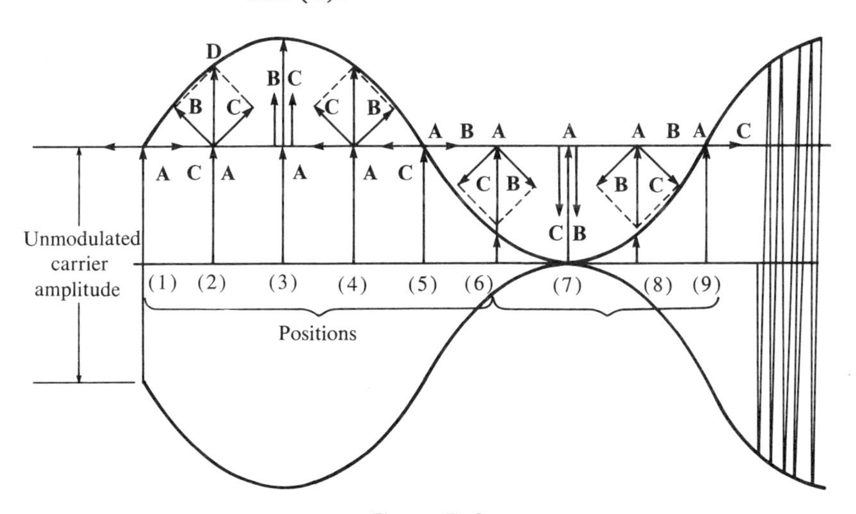

Figure 7–3

Two Phase

Color information can be conveyed by two chrominance signals and a luminance signal. A technique has been developed that permits the transmission of an additional carrier within the same space occupied by the luminance channel. The two chrominance signals are transmitted on this single carrier. This requires the combination of the two chrominance signals.

To combine the two chrominance signals, a technique known as two-phase modulation is used. In ordinary amplitude modulation, the instantaneous amplitude of the modulated carrier varies, but its phase remains constant. Only one separate signal may be sent on a modulated carrier in this manner. In two-phase modulation, two carriers, of the same frequency but 90° apart in phase, are each independently modulated by one of the two chrominance signals. The two carriers are then added, producing a single carrier in which both instantaneous amplitude and phase vary. Two separate signals may be sent on a modulated carrier in this manner. The two separate signals may be independently recovered.

Figure 7–4 illustrates the two-phase modulation technique. The frequency of the carrier used for transmitting the color information is approximately 3.58 MHz. It is transmitted within the 6-MHz spectrum of the TV station, and is called the color subcarrier. This 3.58-MHz subcarrier is applied directly to the A modulator, through a 90° phase-shifting device (such as a tuned transformer), and to the B modulator. As shown, one chrominance signal is fed to the A modulator. The other is fed to the B modulator. Briefly, each modulator could consist of two tubes or transistors operating with opposite polarity inputs and a common output. This type of operation (suppressed carrier) cancels the individual chrominance signal and subcarrier components that feed the

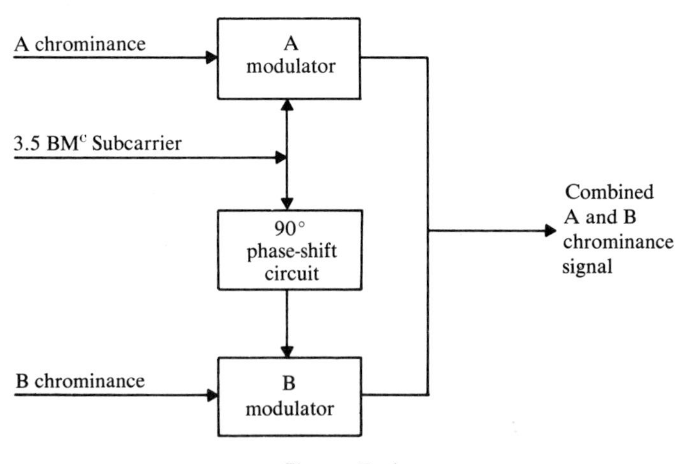

Figure 7–4

modulators. The only signals appearing in the common plate, or output circuit, are the side-band signals.

A vectorial representation of the two-phase modulation technique is shown in Figure 7–5. The output of modulator A is shown in Figure 7–5(a). Note that the side bands produced by amplitude modulation are drawn in a conventional manner. The carrier, which is suppressed, is indicated by a dashed line. The resultant of the side bands, which always falls in phase with the carrier, represents the output of the modulator at any point in time.

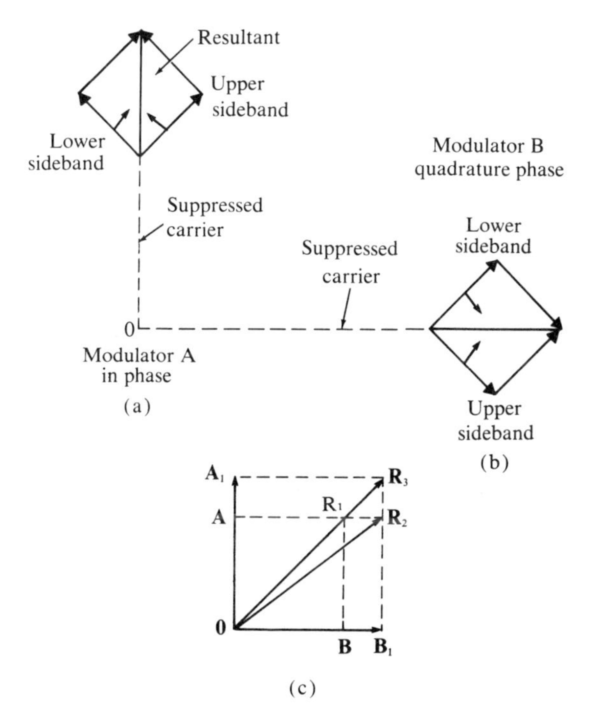

Figure 7–5

By combining the outputs of these two balanced modulators and varying this amplitude and polarity separately, we can produce a sine-wave signal whose phase will fall anywhere within the 360° of our color circle. Figure 7–6 illustrates this.

The outputs of balanced modulator #1 and balanced modulator #2 are combined in an adder to give us a resultant output. The four illustrations show how the phase of the resultant can be made to appear in the four quadrants of our color wheel.

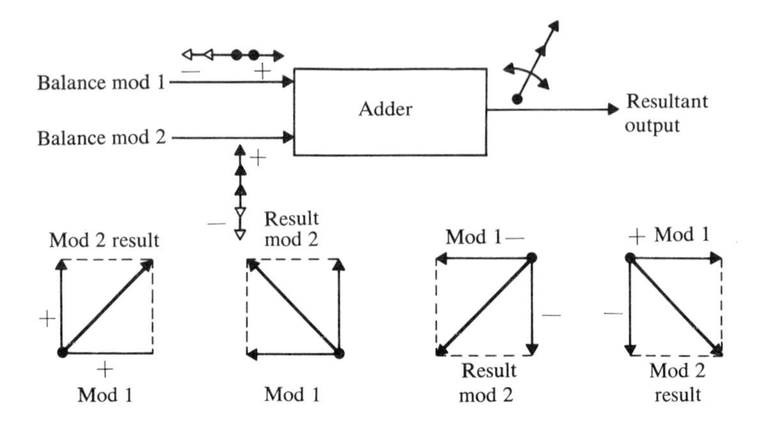

Figure 7–6

Figure 7–7 depicts the information just described in a block diagram.

To summarize, we now have all the components necessary for transmission of color information, excluding the vertical and horizontal sync pulses (Figure 7–8). These components include the $R - Y$ signal, the $B - Y$ signal, the chrominance signal, the correct phase of $E_R - Ey$ with reference to $E_B - Ey$ signals, and the 3.58 color synchronizing pulse (color-burst signal).

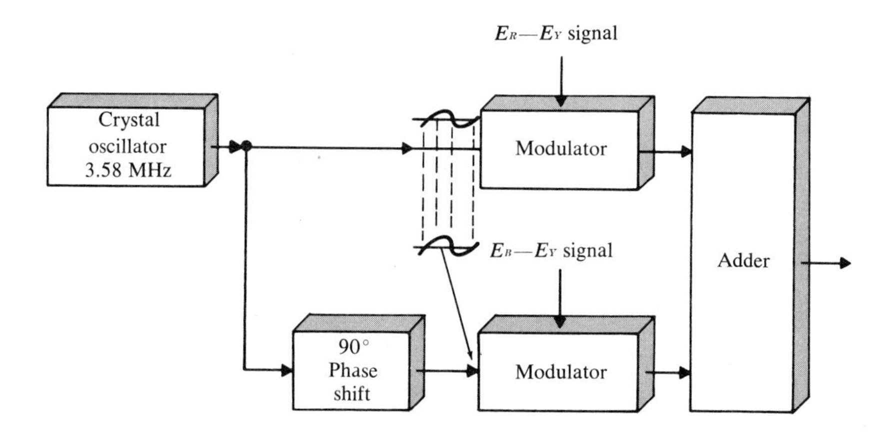

Figure 7–7

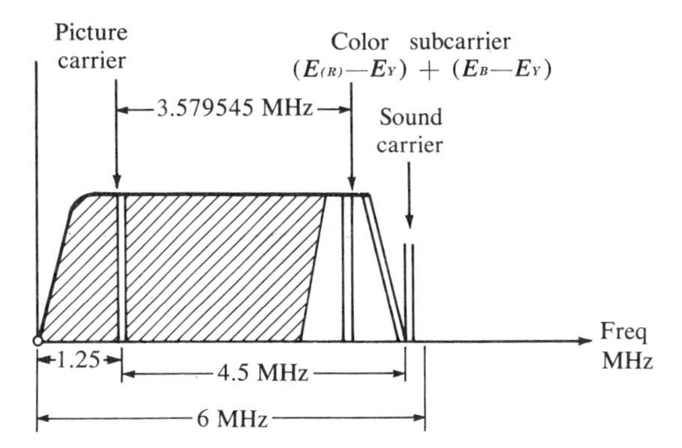
Picture carrier
Color subcarrier
$(E_{(R)} - E_Y) + (E_B - E_Y)$
3.579545 MHz
Sound carrier
Freq MHz
1.25
4.5 MHz
6 MHz

Figure 7–8

questions

1. What is modulation?
2. Describe the difference between amplitude modulation and frequency modulation.
3. What is suppressed-carrier modulation?
4. Demonstrate by the use of vectors what 100% modulation would look like and explain.
5. What analogy could be used to demonstrate that the carrier appears to stand still during the process of modulation?
6. What type of modulation is used to combine the two chrominance signals?
7. What is the frequency of the color-burst signal?
8. The transmitted color information consists of the following: ________, ________, ________, ________, ________, ________.
9. The combination of two chrominance signals uses the technique known as ________.
10. Describe two-phase modulation.

8 color transmission and reception

Chapter 2 described the basics of transmission and reception of black-and-white signals. This chapter will discuss the functional steps taken to add the color (chrominance) information to the black-and-white signal (brightness). The audio and synchronizing functions will be omitted in these descriptions.

Transmission of a Red Scene

Figure 8–1 (see also front endpapers) shows the transmitter signals for a red-bar transmission. When all three cameras focus on a red image, the green and blue cameras have no output, since there is only red color in the image. The red camera, however, will have one unit of output which is fed directly to the red adder and through the voltage divider to

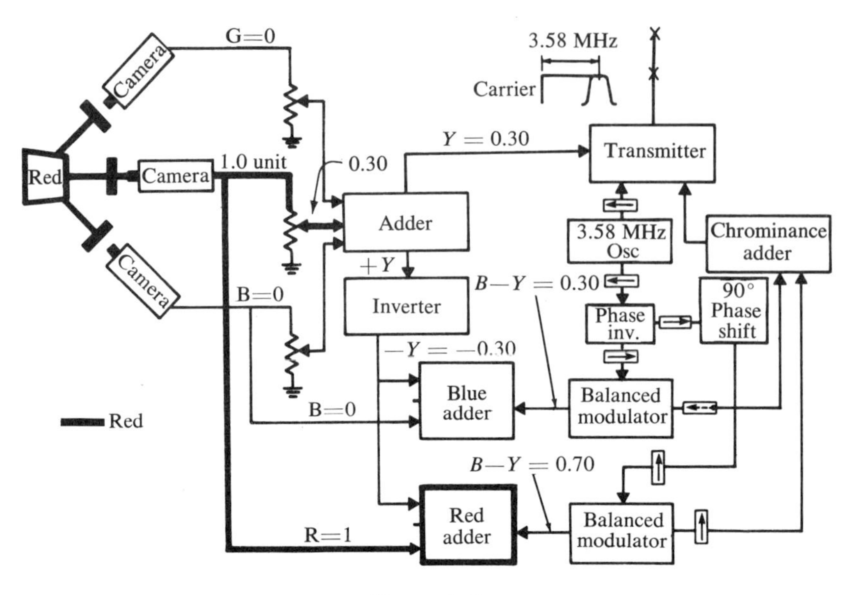

Figure 8–1

pick off 30% brightness level for compatible black-and-white transmission. Since the green and blue signals are zero, they will contribute nothing to the brightness-adder output—our black-and-white, or brightness, signal will be a 30% value. This 30% brightness signal is inverted (changed from a positive signal to a negative signal) and is applied to the blue and red adders as a −30% signal. In the red adder, we have a +100% signal from the red camera and a −30% signal from the black-and-white inverter. These two signals, added algebraically, yield a positive 70% signal (100% − 30% = 70%). Since the brightness signal has been designated as the *Y* signal, the output of the red adder is designated an *R*− *Y* signal. It is the red information, minus the *Y* (or brightness) information.

The inputs of the blue adder are the −30% black-and-white signal and the zero signal from the blue camera, since there is no blue-camera output on a red scene.

The −30% signal out of the blue adder modulates the blue-balanced modulator which gives the output shown by the dotted vector in the 180° position. The red-balanced modulator is modulated by the +70% (*R* − *Y*) signal. Its output is indicated by the solid vector in the 90° position.

The outputs from both balanced modulators are sent to the chrominance adder where they form the resultant signal, as shown, and modulate the video carrier. The color signal will cause a side band of the video carrier at the same frequency used to excite the balanced modulators. This frequency is 3.58 MHz. Since it appears at the high-frequency end of the video response band, it will result in low visibility information in the black-and-white transmission. The black-and-white signal modulates the transmitter as a full bandwidth signal since all fine detail is transmitted in the black-and-white signal.

Reception of a Red Scene

Now let us follow red-bar transmission through a color receiver. In Figure 8–2 (see also back endpapers), we have the block diagram of our color receiver. It shows the tuner, IF, and detector, with the expected bandwidth response curves.

The tuner bandwidth must be wide enough to pass the total transmitted radio frequency (RF) signal. This includes the 4.2-MHz wide video carrier (including color subcarrier) and the sound carrier. The sound carrier is 4.5 MHz higher in frequency than the video carrier.

A local oscillator in the tuner mixes with the amplified RF signal and produces an intermediate frequency (IF) which is amplified at a fixed frequency. The IF amplifier has a bandwidth of 3.58 MHz which places the video-IF carrier at 50% on one side of the curve, the color subcarrier at 50% on the other. The IF beat frequency (the difference between the video carrier and the audio carrier) appears 4.5 MHz lower in frequency than the IF carrier and is sharply attenuated (reduced) to prevent interference in the picture. This does not affect the sound information since it has a separate IF amplifier.

After amplification, the IF signal is detected to recover the complete video signal (including color subcarrier). This will be a wide-band signal with frequencies from 30 Hz to over 4 MHz. (Notice the similarity to black-and-white.) This signal takes two paths (Figure 8–2).

The first path travels through the video amplifier. This amplifies the black-and-white portions of the signal. At the cathodes (electron-emitting elements) of the three guns of our color picture tube, we recover a 30% brightness signal that contains all of the fine detail and brightness information. Since this is applied to all three cathodes, all three guns have a 30% "on" signal. Let us leave this signal here for a moment to go back and follow the other path taken by the detected video signal.

Let us first consider the color-IF amplifier. It is tuned to a center frequency of 3.58 MHz, with a total bandwidth of 1.0 MHz. This amplifier is sensitive only to color subcarrier frequencies and will exclude all brightness information. The output of the color-IF amplifier is the amplified color signal which contains hue and saturation information. This is applied to the grids of the two demodulators (which recover color information).

Part of the amplified signal from the color-IF amplifier is channeled to the color-sync amplifier, which amplifies only during horizontal retrace time (the only time that the color sync is present). The color-sync burst is amplified and sent into a phase detector which compares its phase and frequency with the local crystal oscillator. The phase detector then provides correction voltages to the crystal oscillator, maintaining phase and frequency stability. The 3.58-MHz oscillator has two outputs. The 90° phase output excites the red demodulation. The 0° phase output excites the blue demodulator.

The demodulators will have voltage outputs determined by the phase of the color signal being injected into them. The red demodulator will have maximum output in the positive direction when the color signal is in the same phase as the exciting voltage. It will have maximum negative output when the color signal is 180° out of phase with the exciting voltage. The blue demodulator will have maximum positive output when the incoming color signal is in the same phase as its exciting voltage. It will have maximum negative voltage output when the incoming color signal is 180° out of phase with the exciting voltage.

The output from the red demodulator is fed into the $R - Y$ amplifier and appears as a positive 70% signal. It is connected to the red grid of the color tube. The output of the blue demodulator is fed into the $B - Y$ amplifier and appears as a −30% signal. It is applied to the grid of the blue gun in the color tube. When we described the color wheel in chapter 4, we determined that any color in the color wheel could be expressed and defined in terms of $B - Y$ and $R - Y$ directions. We also learned that in order to recover the $G - Y$ signal at the receiver, it was necessary to take certain amounts of the $R - Y$ and $B - Y$ signals. The amounts needed here are −51% $(R - Y)$ and −19% $(B - Y)$. This results in the $G - Y$ signal. By putting this combination into the $G - Y$ amplifier, we come out with a −30% signal (in this case) for the signal red. It is applied to the grid of the green gun of the color tube.

Examining the picture tube, we see that on the cathodes of all three guns we have a +30% brightness, or "on", signal. On the grid of the red gun we have a 70% "on" signal which adds to the 30% brightness signal. The red gun is turned on 100%. On the green gun, we have a −30% signal on the grid and a +30% signal on the cathode. These cancel, yielding no signal on the green gun. On the blue grid, we have a −30% signal and on the cathode we have a +30% signal. These two also cancel, giving the blue gun a zero signal. The red gun, then, is turned on 100%. The green and blue guns are turned off, giving only a red raster which reproduces the original scene.

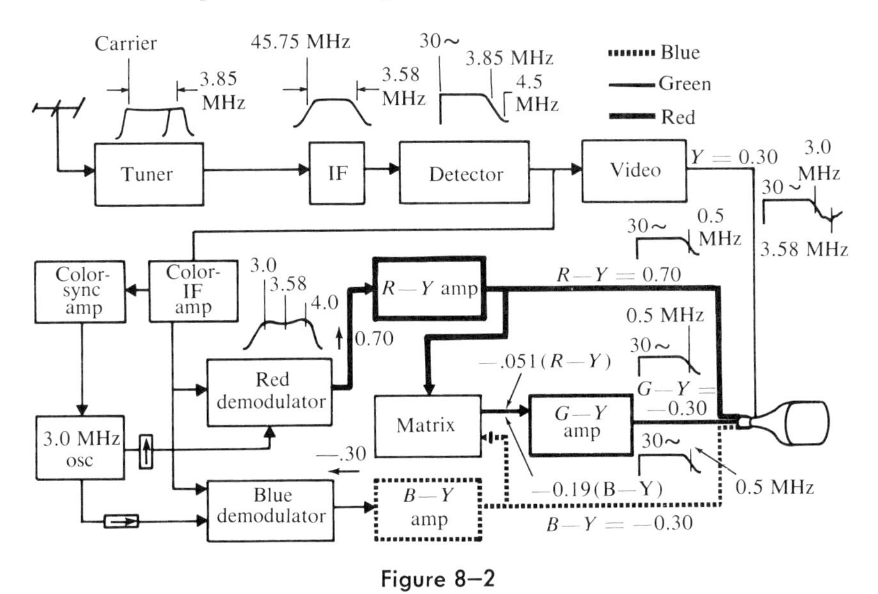

Figure 8–2

Transmission of a Green Scene

Figure 8–3 shows the signals transmitted for a green bar. With this condition, only the green camera will have output. Fifty-nine percent of this signal is fed into the black-and-white signal adder. Because green is only 59% as bright as white to the eye, this is the brightness level transmitted in the black-and-white channel.

The red and blue cameras have no output, so no percentage of these signals is included in the brightness signal. This zero voltage is added to the inverted black-and-white signal, giving both colors a −59% output signal. The outputs of both balanced modulators will have a negative signal output, which when combined in the chrominance adder, will modulate the transmitter with the phase shown. The green-bar transmitter then will transmit the black-and-white signal with a brightness level of 59%, the color subcarrier with a phase of approximately 241°, and the synchronizing information.

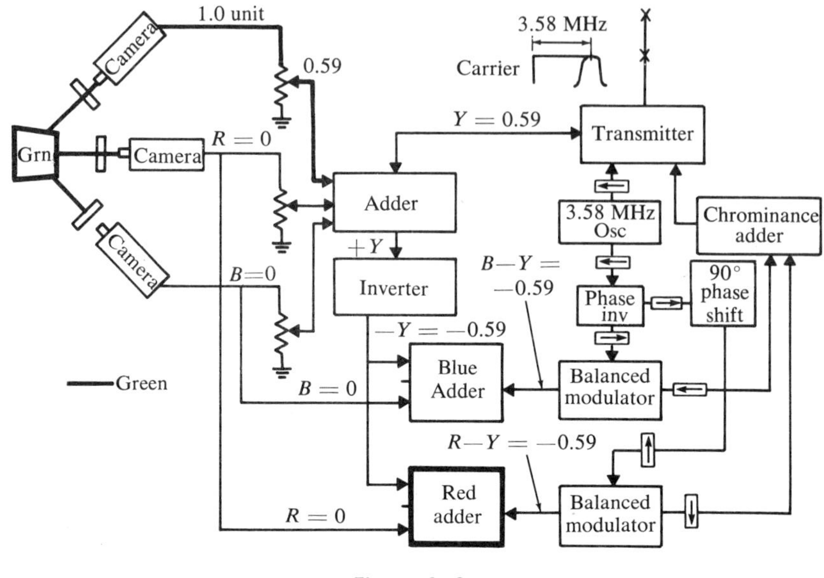

Figure 8–3

Reception of a Green Scene

In the color receiver (Figure 8–4—see also back endpapers) the output of the detector contains all of the video-modulating information. The output is separated into its various paths. At the output of the video amplifier we find the black-and-white signal representing a 59% brightness level. This is applied to the cathodes of all three guns.

The color-IF amplifiers amplify the color subcarrier. Because of the subcarrier's bandpass response, the IF amplifiers exclude all brightness information and send this signal to the grid of both demodulators.

The color-sync amplifier amplifies only the sync information and controls the phase and frequency of the local 3.58-MHz oscillator. The outputs of both the red and the blue demodulators will be a signal representing negative voltage. They will appear as -59% signals in the outputs of both the $R - Y$ and $B - Y$ amplifiers. These signals are applied to the grid of the red and blue guns.

Again, by taking the proper percentages of the $R - Y$ and $B - Y$ signals, we develop the $G - Y$ signal. The $G - Y$ signal is $+11\%$. It is applied to the grid of the green gun and, combined with the $+59\%$ signal on the cathode, turns on the green gun 100%. Since the red and blue guns each have a -59% signal on their grids and a $+59\%$ on their cathodes, signals cancel and the red and blue guns are turned off. The green gun is turned on 100%, producing a green raster which reproduces the original scene.

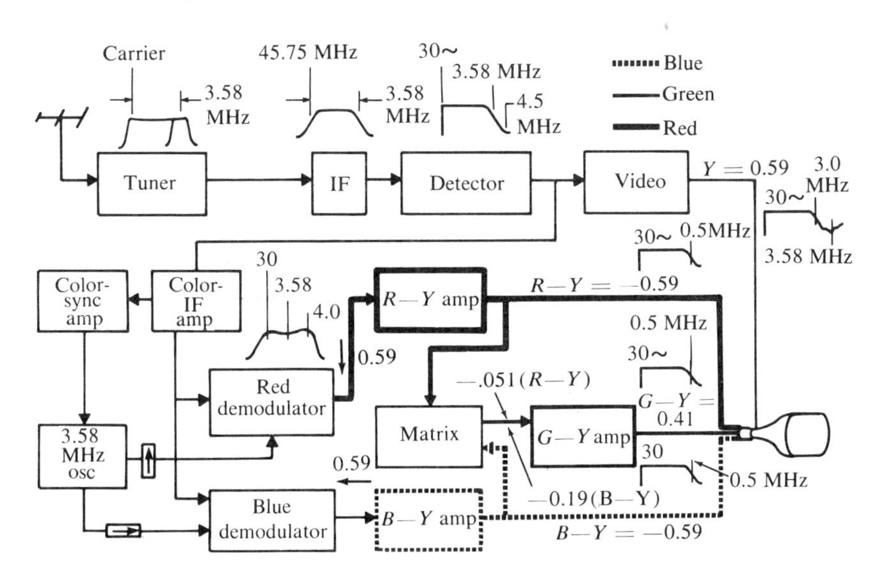

Figure 8–4

Transmission of a White Scene

Figure 8–5 (see also back endpapers) shows the transmission of a white scene. Since white is all colors of light combined, each of the three cameras will have maximum, or one unit, output. The outputs of all three cameras are broken down in the proportions of brightness seen by the human eye. They are combined in the adder to produce the black-and-white, or brightness, signal. Again, these proportions are 59% green, 30% red, and 11% blue. Totaled, they equal 100%.

The total output of the red camera is fed into the red adder and the total output of the blue camera is fed into the blue adder. Since all three cameras put out 1.0 unit and we indicate the brightness level of the scene in terms of certain percentages of three colors, the black-and-white signal equals 1.0 unit ($0.59G + 0.30R + 0.11B = 1.0$). This is the maximum brightness that will appear in the transmission. This black-and-white signal is inverted, becoming a -1.0 unit, and is inserted into the red and blue adders.

In the blue adder, we have a +1.0 signal (from the blue camera) and a −1.0 signal (from the black-and-white inverter), yielding zero. The output of the blue adder is zero.

The color signal in the red adder is +1.0 unit. The inverted black-and-white signal is −1.0 unit. These cancel and the output from the red adder is also zero.

Since there is no signal input to either of the balance modulators, no oscillator signal will go out. On a transmission that contains only black-and-white information there will be no subcarrier or color signal formed.

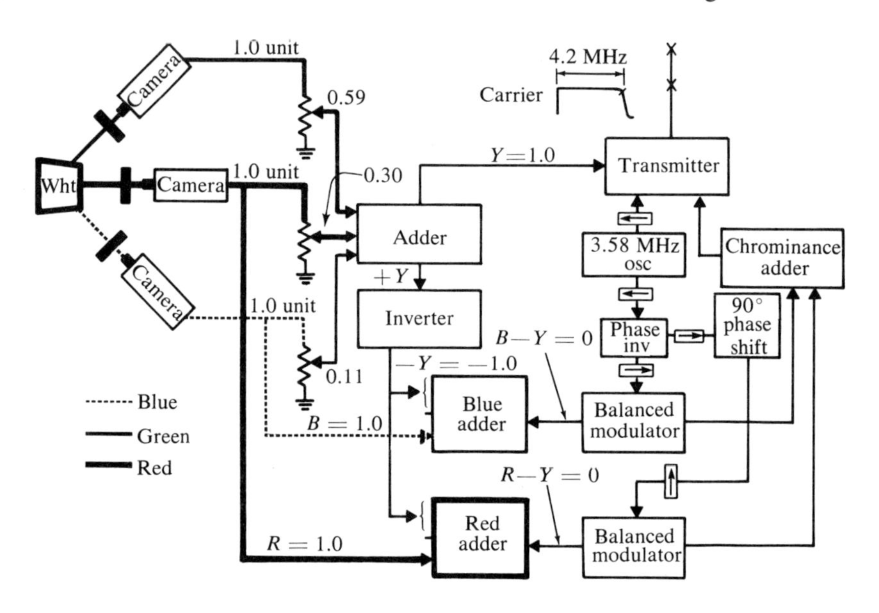

Figure 8–5

Reception of a White Scene

Figure 8–5 shows how this signal will be received in a color receiver. The tuner accepts the total signal, both sound and video. Here it is converted to an IF and amplified with the proper bandwidth. From the IF, the signal goes into the detector, where the video information is removed. The bandpass response of the detector ranges from 30 Hz to 40 MHz, dropping off sharply at 4.5 MHz. From the detector, the video signal goes to the video amplifier where it is amplified and emerges as a signal of 1.0 unit amplitude. It is applied to the three cathodes of the color tube.

Since there is no color information in this transmission (only black-and-white information), there will be no signals through the color circuits. The outputs of the three color amplifiers will be zero. The three color grids have no signal, but the cathodes of the three guns have a 1.0 unit, or a 100%, signal. Each of the guns is turned on 100% giving a white raster and duplicating the original scene.

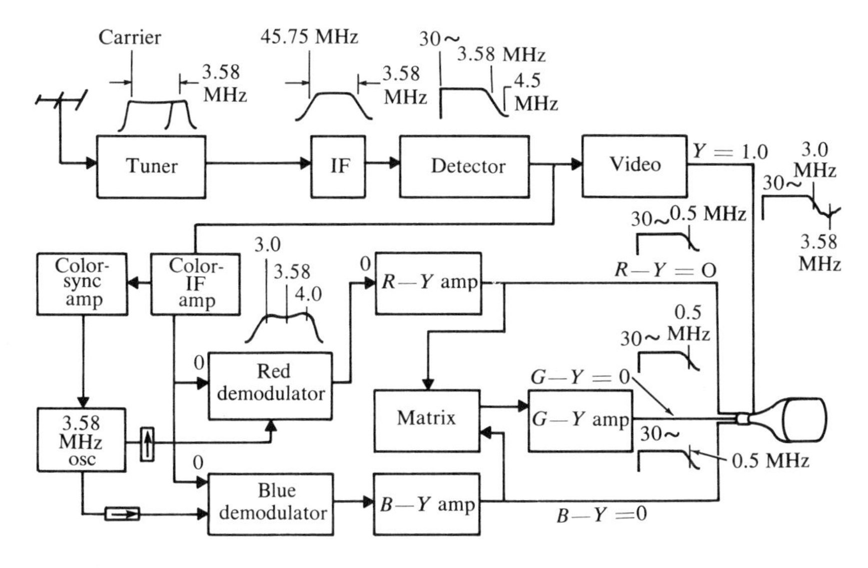

Figure 8–6

questions

1. When all three cameras focus on a red scene, the green camera and blue camera will have 0.59G and 0.11B outputs.

   True _____ False _____

2. A color TV camera uses—

   (a) one camera tube with three filters

   (b) three camera tubes

   (c) one camera tube and interleaving

3. Identify the section of a color TV receiver in which the Y and chroma signals are separated.

4. Explain the function of the adder as used at the transmitter.

5. Why is it unnecessary to separately generate a $G - Y$ signal at the transmitter?

6. What is the condition of the three cathodes in the picture tube when a red signal is being received?

7. What is the condition of the three cathodes in the picture tube when a green signal is being received?

8. What is the condition of the three cathodes in the picture tube when a white signal is being received?

9. A blue scene was not described. Look at Figure 8–1 and in place of a red scene describe the outputs for a blue scene.

10. Use Figure 8–2 and insert inputs and outputs for a blue scene. Describe in detail.

9 color cathode-ray tube

As high-quality audio is to stereo, high quality picture is to television. Picture quality is only as good as the ability to reproduce it. The picture tube is one of the most important elements in that process.

Black and White Gun Structure

Figure 9–1 shows a cross section of a black-and-white picture tube. When the cathode of the electron gun is heated, a stream of electrons are emitted and attracted toward the front of the picture tube. The arrangement of the electron gun is shown in Figure 9–2.

The potential difference between the control grid and the cathode determines the amount of beam current and, consequently, the brightness of the tube. The screen grid has a positive potential. It accelerates the electron beam into the focus element which forms the stream of electrons so that it strikes the phosphor screen at a small, concentrated spot. The voltage applied to the focus element is positive and is usually adjustable. The accelerating anode (electrode toward which electrons flow) increases the beam's velocity and provides sufficient energy to fully illuminate the phosphor screen. The beam of electrons is caused to scan the tube screen by means of an externally mounted deflection yoke. As current flows through the coils, magnetic fields are produced within the neck of the tube. The electron beam passing through this field is deflected in both a vertical and a horizontal direction to produce a raster.

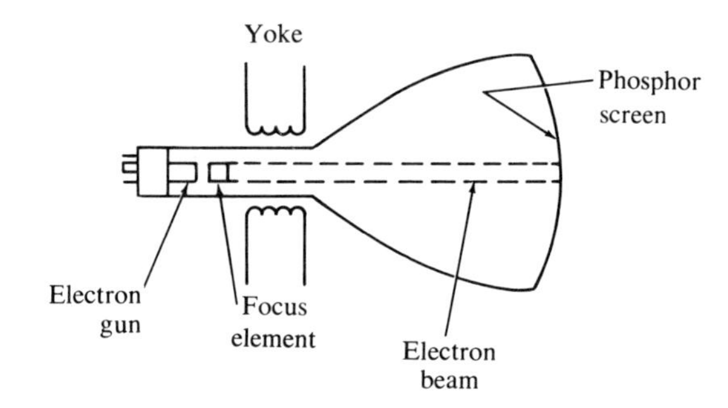

Figure 9–1

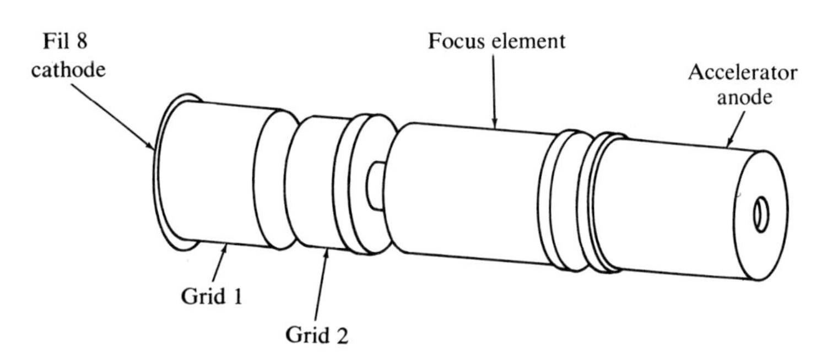

Figure 9–2

Color Tube Gun Structure

The color cathode-ray tube, as used in most receivers, has three electron-gun assemblies. Each gun in the three-gun assembly is quite similar to the electron gun used in a black-and-white picture tube. The three guns are arranged 120° apart, around the center axis of the tube.

All three guns are aimed at a spot at the center of the screen. Figure 9–3 shows the relative position of each of the three guns.

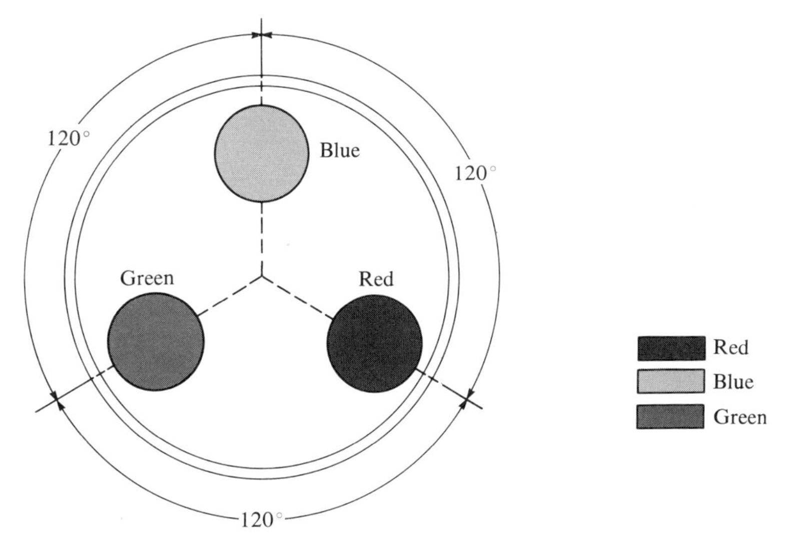

Figure 9–3

Shadow Mask

The screen of a tri-gun color tube is considerably different from the phosphor screen in a black-and-white tube. Instead of a single phosphor coating on the face of the tube, three different phosphors, capable of emitting three different colors of light, are used. The three phosphors are arranged as a series of dots. Directly behind the faceplate is a shadow mask which contains one hole for each group of three phosphor dots on the faceplate. The streams of electrons from the guns must pass through the holes in the shadow mask in order to strike the phosphor material and produce light.

The three electron beams must be controlled so that each beam strikes only the phosphor dots of one color. The shadow mask makes this possible. The individual dots of phosphor material are placed on the faceplate of the picture tube as shown in Figure 9–4. The three dots are arranged in a triangular pattern known as a *triad*. The triads, or groups of three dots, cover the entire faceplate. The shadow mask contains about 500,000 holes with one hole behind each triad. The mask is positioned so that each hole is located directly in front of the center of a triad.

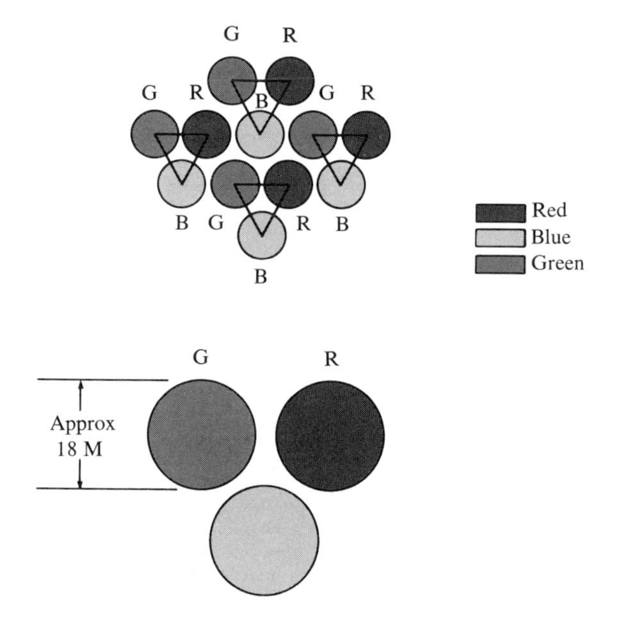

Figure 9–4

Figure 9–5 shows the relative position of the shadow mask to the phosphor triads. Each opening in the shadow mask is directly over the center of each triad. When viewing the opening from a point perpendicular to any one hole in the mask, we see a small portion of each color phosphor.

Because the three guns do not have the exact same point of origin, the electron beams approach the mask from three different angles, none of which is perpendicular to the hole. If the approach angles are correct, each electron beam will strike only the proper color. This principle is demonstrated in Figure 9–6.

Placement of Color Phosphors

An explanation of how the color phosphors are placed on the faceplate will help in understanding the shadow mask principle. Initially, a faceplate and shadow mask are selected. The mask is mounted into the faceplate in such a manner that it can be removed and later reinstalled in exactly the same position. The same faceplate and mask remain to-

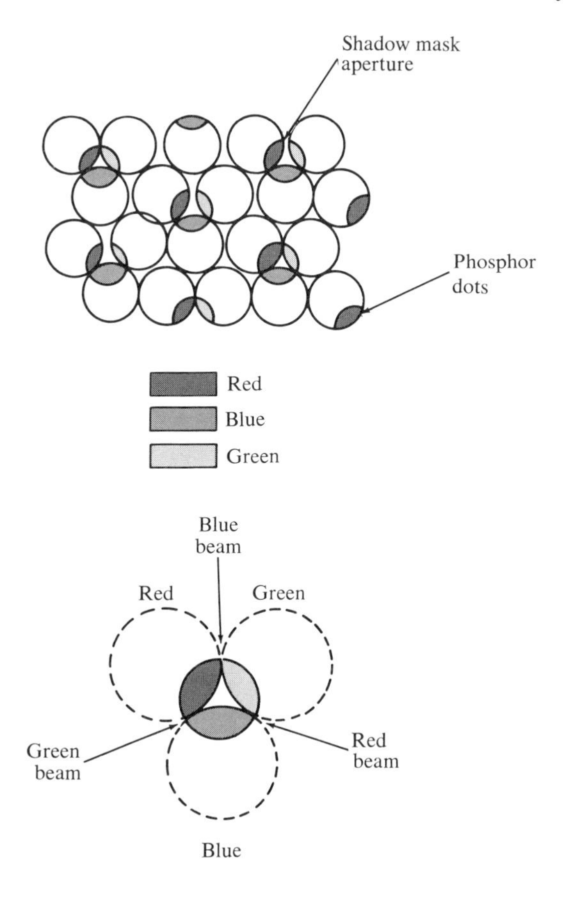

Figure 9–5

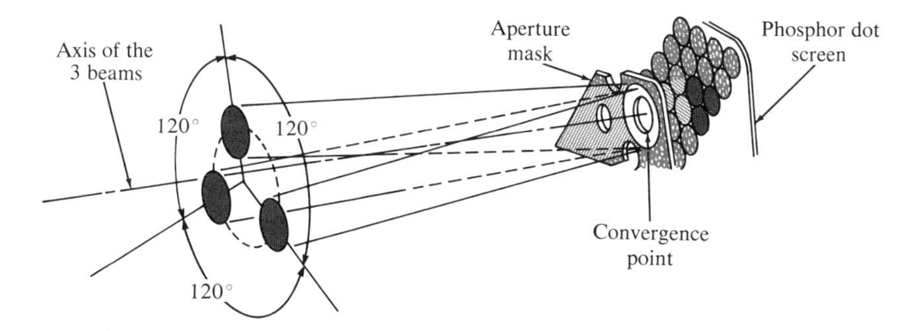

Figure 9–6

gether during the manufacturing process. The faceplate is covered with an even coating of photosensitive green phosphor material. The shadow mask is then placed accurately in the faceplate in a special jig. A point-size light source is placed in the position the green gun would occupy in the tube. It floods the whole shadow mask with light. When the light shines through the holes in the shadow mask and hits the phosphor coating, the phosphor material becomes hardened and sticks to the glass. The faceplate is then sprayed with water, removing the phosphor that was not exposed to light. The exposed wafers of phosphor material are left adhering to the glass.

An even coating of photosensitive blue phosphor material is then put over the faceplate, which now includes the hardened, green phosphor wafers. The shadow mask is accurately put back into position and the light source is moved to the position the blue gun will occupy in the finished tube. The blue phosphor coating hardens when the light strikes it. The light strikes adjacent to the green phosphor dots which are already present. At the completion of the exposure, the faceplate is again washed, removing the unexposed blue phosphor material. Blue phosphor wafers are left where the light source has hardened them.

Finally, the faceplate is coated with a photosensitive red phosphor material and the shadow mask is again accurately positioned. The light source is now moved to the position that the red gun will occupy in the tube. It shines through the shadow mask holes, causing red phosphor wafers to be exposed and hardened.

The shadow mask is then removed again and the unexposed red phosphor material is washed away. We now have a precise arrangement of the three-color phosphor dots in front of the shadow mask which will be sealed into the finished tube.

After the phosphors for each of the three colors are deposited, we can alternately place our eye at the point where each gun will be located in the finished tube and see phosphors of only one color. Since we have controlled the position of the light source that exposed the individual dots, we can install the electron guns in the same positions and each gun will strike phosphor dots of one color.

Purity Device, Purity Adjustment

A black-and-white picture tube requires a deflection system with provisions for centering and focus controls to provide control of the electron beam. In order for the color picture tube to correctly reproduce color pictures, the individual electron beams must light phosphors of only one color. Each electron beam is then capable of producing a pure field of either red, blue, or green.

In order for each beam to light only the proper phosphor, it is of extreme importance that the approach angle of the electron beams to the shadow mask be correct. The correct angle is the one that duplicates the trajectory of the light source that formed the dots. An incorrect approach angle will result in an electron beam lighting phosphors other than the correct color. Were this to happen, the tube would be incapable of reproducing pure colors.

Figure 9–7 shows the effect of an improper approach angle. Each electron beam is only partially hitting the correct phosphors. The remaining portion of the beam strikes other phosphors of the wrong color. In Figure 9–7, the electron beams are too high and are not centered on the triad. The arrows indicate the direction of travel needed for correction.

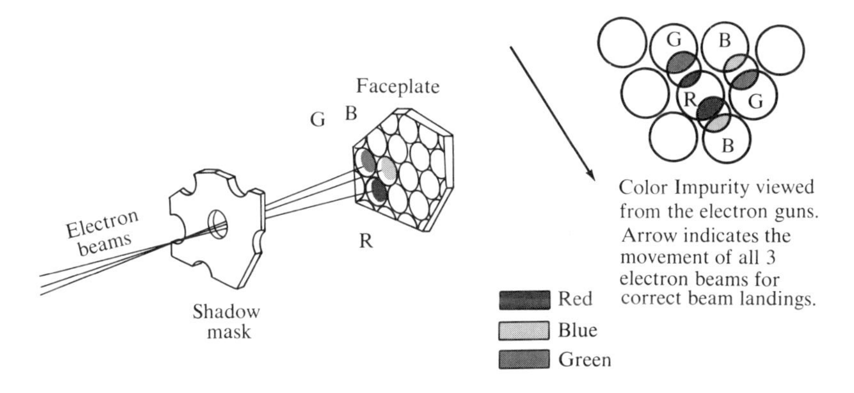

Figure 9–7

It is not always possible to place the electron guns in the tube accurately enough (in relation to the shadow mask) to always have a perfect approach angle. Therefore, some means must be provided to correct any approach-angle error of the electron beam. The corrective action must be made before the three beams pass through the shadow mask.

An electron stream passing through a magnetic field will be deflected at right angles to the field. The amount of deflection will depend on the strength of the field. By placing a magnet that has adjustable strength and direction of field on the neck of the tube, we can control the approach angle of the electron beams to the mask.

A color purity device that supplies the proper field is shown in Figure 9–8. The purity-magnet assembly consists of two similar ring magnets. Each magnet has a square and a round tab. The square tab in magnet A is the north pole while the round tab in magnet B is the south pole. When the magnets are assembled with like tabs together as shown in C, the two fields will cancel, since they are in opposition. The purity magnets may be moved independently or rotated together.

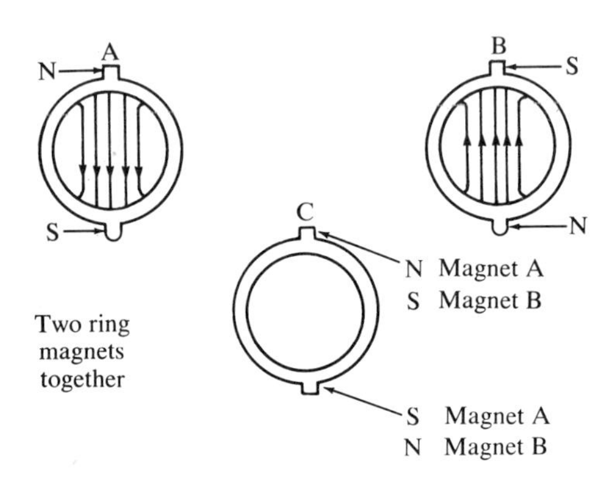

Figure 9–8

Figure 9–9 shows the action of the purity device as it is adjusted.

Figure 9–9(a) shows the purity device with tabs together. With the tabs so adjusted, there will be no movement of the three electron beams. They will pass through the device as if it were not present.

Figure 9–9(b) shows the tabs separated so that a magnetic field is produced. The electron beams passing through the purity device will be deflected at right angles to the flux lines. The beam movement will be in the direction indicated by the arrow. Notice all three beams move in the same direction, by the same amount.

Figure 9–9(c) shows the same tab spacing, but the entire assembly has been rotated 120°. Since the tab spacing has remained constant, the strength of the field is the same as in 9–9(b). The device has been rotated so the direction of the flux lines has changed. The beams will be deflected in the direction indicated by the arrow.

Figure 9–9(d) has the same flux direction as 9–9(c), but the tabs are open to produce a stronger field. The deflection direction is the same, but the beams have been deflected a greater amount by the stronger field. By adjusting the tab opening, we can adjust the amount of beam deflection. By rotating the assembly, we can adjust the direction in which the beams are deflected. Thereby, we can position the three beams in the tube neck so that they approach the shadow mask at the proper angle or appear to originate from the same point as the original light source used to place the phosphors on the screen.

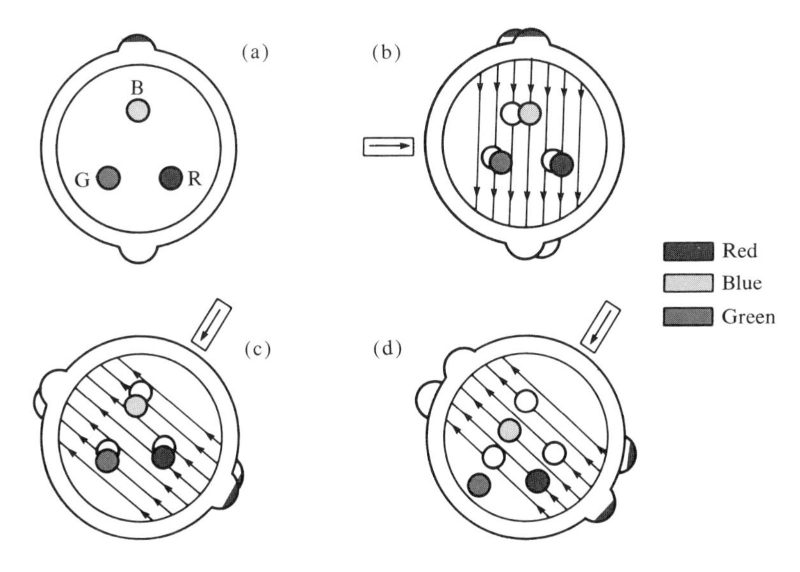

Figure 9–9

We can apply the rules of beam deflection through a field by using the left-hand rule described in chapter 3 on magnetism. Lay your left hand flat [adjacent to Figure 9–9(b)], with the index finger pointing in the direction of the arrows (flux lines). The middle finger points up from page indicating current flow (electron beam). The thumb shows direction of beam movement. This same method can be followed by applying a similar technique to Figure 9–9(c) to verify the beam movement shown.

The purity device is usually located around the neck of the picture tube before the beams are deflected by the yoke. Figure 9–10 shows a side view of how adjustment of the purity device can change the apparent point of origin of the three electron beams.

Figure 9–10(a) shows the purity device with the tabs open and the direction of its field such that the three beams will be deflected upward. In 9–10(b) the device is rotated 180° so the three beams are deflected downward. The magnetic purity device is adjusted to produce pure fields as viewed at the center of the screen.

For outer areas of the screen, it is important that the electrical center, or deflection center, of the yoke be correctly located in order to main-

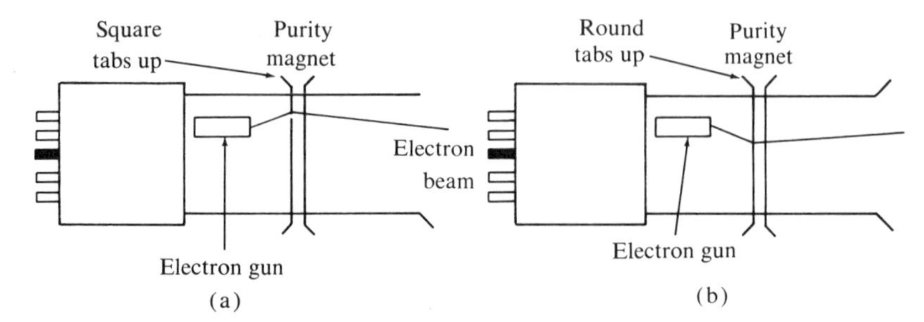

Figure 9–10

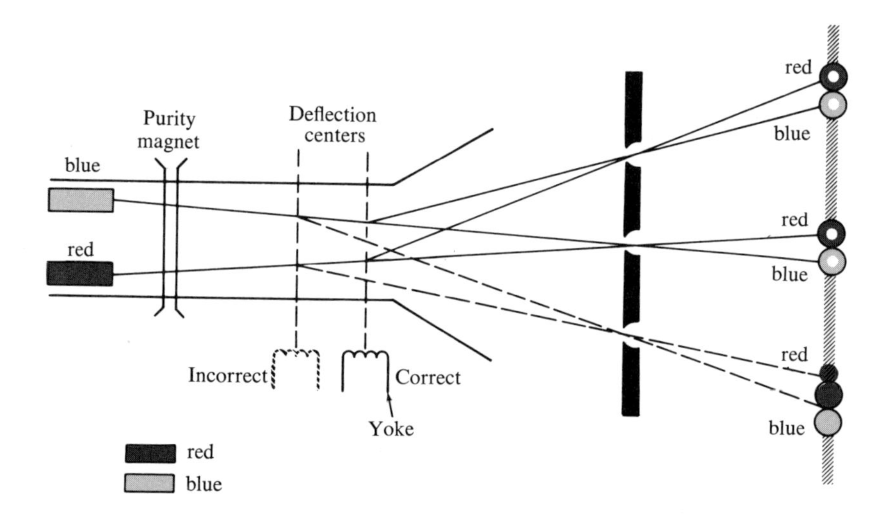

Figure 9–11

tain the proper approach angle to the screen. Purity adjustments for areas outside the center of the screen are made by sliding the yoke along the neck of the tube.

Shown in Figure 9–11 are both the effect of mounting the yoke with the deflection center located properly, and the effect of mounting in a second, and incorrect, position, located too far to the rear. When the deflection center is correct, the electron beams will leave the yoke field, pass through an opening in the shadow mask, and illuminate the correct phosphors (top of Figure 9–11). If the yoke is moved toward the rear of the tube neck (moving from the correct deflection center), the electron beams will pass through the corresponding hole at the bottom of the mask at a different angle, missing the correct phosphors.

Figure 9–12 shows the direction of movement of beam landing as the yoke is moved forward and backward. As the yoke is moved along the neck of the color tube, the impact point will move along the lines drawn through the triads shown in Figure 9–12.

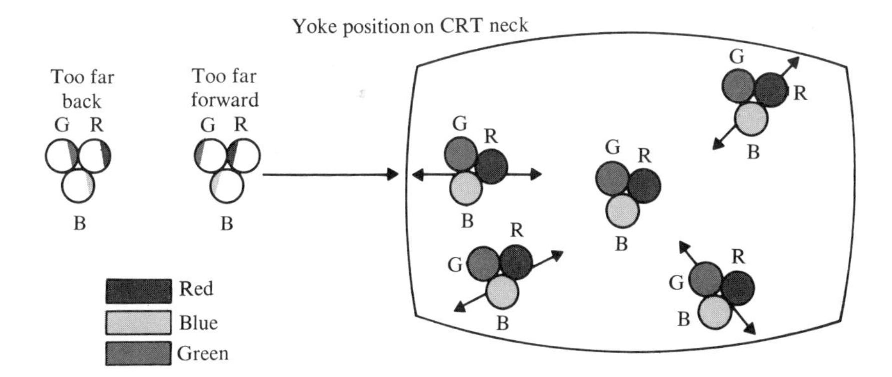

Figure 9–12

Notice that the area at the center of the screen does not change as the yoke is moved (Figure 9–13—see also front endpapers). Purity adjustments are started with the yoke moved as far to the rear as possible. The magnetic purity device is adjusted for correct strength and polarity to produce a pure field at the center of the tube. The green and blue guns

Figure 9–13

are turned off, using the screen-grid controls for these adjustments. Only the red gun is left operating. The red field is used because the red phosphors are the least efficient and impurity can be seen more easily. If the beam is partially hitting another color phosphor, its more efficient light output would make the error readily apparent. After the purity device is adjusted to produce a pure red field at the center of the screen, the yoke is slowly moved forward until a pure red field is produced at the edges of the screen. The green and blue fields are then individually checked for purity by turning down the red screen control and turning on the other two screen controls in turn.

questions

1. The black-and-white tube has one gun. How many guns do present-day color tubes have?
2. The gun structures are placed _____ degrees apart.
3. The three color dots form what kind of geometric form?
4. The triangle of phosphor dots is called a __________ of dots.
5. Explain the meaning of *purity*.
6. The __________ ring is used to provide proper beam landing to produce proper color of the three fields.
7. The yoke position affects the (center), (edge) purity.
8. How many phosphor coatings are applied to the face of the picture tube?
9. Why is the shadow mask necessary?
10. Why is the approach angle to the shadow mask of such great importance?
11. Describe the purity-ring construction.
12. How do the purity rings move the electron beam?

10 convergence

To produce a white field on the screen of the picture tube, a mixture of red, green, and blue phosphor illumination must be present. To achieve this mixture, the three rasters must be converged.

It is not only necessary that three pure rasters be produced. They must also be superimposed on one another, at all points on the screen. When the three electron beams, coming from the guns, are positioned in relation to one another so that they cross over at the plane of the aperture mask, the beams are said to be *converged*. When this condition is satisfied, each of the three electron beams produces a raster. The three rasters will be superimposed at all points and are said to be converged.

A crosshatch or dot pattern produced by a generator is used to determine if the three rasters are superimposed, or converged, at all points. The generated crosshatch pattern turns on all three electron beams several times for short intervals as the set is scanned vertically and horizontally. If the three rasters are superimposed at all points, the areas where all three beams are turned on will appear as superimposed vertical and horizontal white lines. If the rasters are not superimposed, red, blue, or green vertical and horizontal lines will be seen.

Convergence-Magnet Assembly

The three electron guns in the color tube are physically aimed at the same spot on the screen, so the three rasters are nearly superimposed or converged at the center of the screen. To allow for slight production variations of the tube, individual beam-aiming adjustments are provided so that the three rasters may be exactly converged at the center of the screen. The beam-aiming device is called a *convergence-magnet assembly* and is shown in Figure 10–1.

The convergence-magnet assembly consists of a horseshoe ferrite (a mixed oxide of iron and other elements used where good magnetic paths are needed for flux lines) core, on top of which is placed a ceramic disc permanent magnet. The polarization of this magnet is indicated by the letters S and N. There are also two coils of wire wound on the horseshoe core, but we will not be concerned with them at this moment since we are only interested in convergence at the center of the screen.

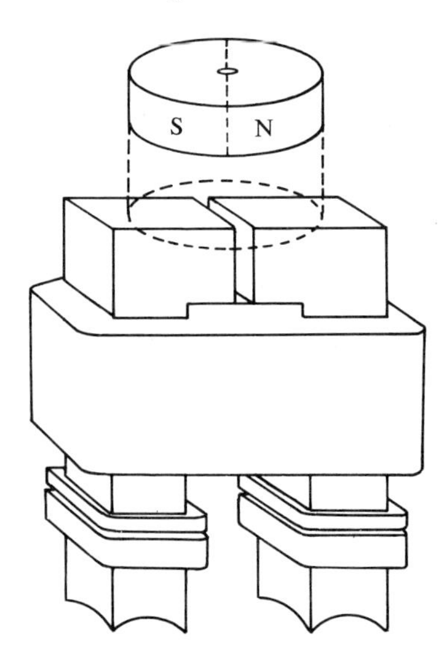

Figure 10–1

Figure 10–2 shows how this convergence-magnet assembly is placed on the tube. Only one is shown in this illustration, but one assembly is placed over each of the three guns.

The gun structure inside the tube is designed so that the electron beam will have to pass between two pole-pieces that correspond to the spacing of the ferrite core of the convergence-magnet assembly. Figure 10–2(a) indicates this arrangement, with the permanent disc magnet turned so that there will be no field across the ends of the core.

Figure 10–2(c) shows the disc magnet turned in the opposite direction, causing the magnetic field to travel toward the right, displacing the raster in the opposite direction.

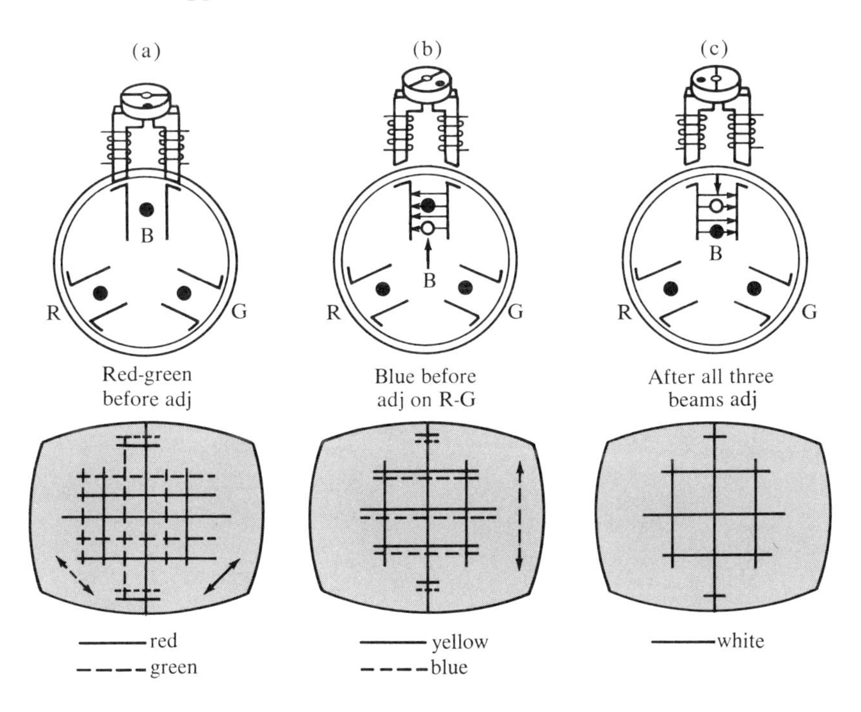

Figure 10–2

Static Convergence

By using three of these convergence-magnet assemblies, we can individually control the radial position of the three rasters on the face of the tube. But again, because of the mechanical limitations, we may have to move the blue beam in a lateral direction in order to make sure that all three beams hit at the same point on the face of the tube.

Blue-Lateral Magnet

To provide this adjustment, an additional pole-piece, the *blue-lateral magnet,* is constructed into the blue-gun assembly. It is shown in Figure 10–3. The *blue-lateral magnet* is a long, cylindrical ceramic magnet that is polarized, as indicated in Figure 10–3(a). As seen here, when the magnetic field goes in one direction, the beam is deflected laterally toward one side of the tube. When the magnetic field is reversed, the beam is deflected to the other side of the tube. The change of direction of the magnetic field is accomplished by rotating the blue-lateral magnet around its own axis. The internal pole-piece in the neck of the tube minimizes interaction with the red and green beams.

In some cases, the blue-lateral magnet may be located on the side of the tube neck rather than on top (as shown in Figure 10–3). Adjustment of the magnet with the device located on the side of the neck will cause red-green and blue to move in opposite directions. This double action will allow blue to be converged with a minimum amount of mag-

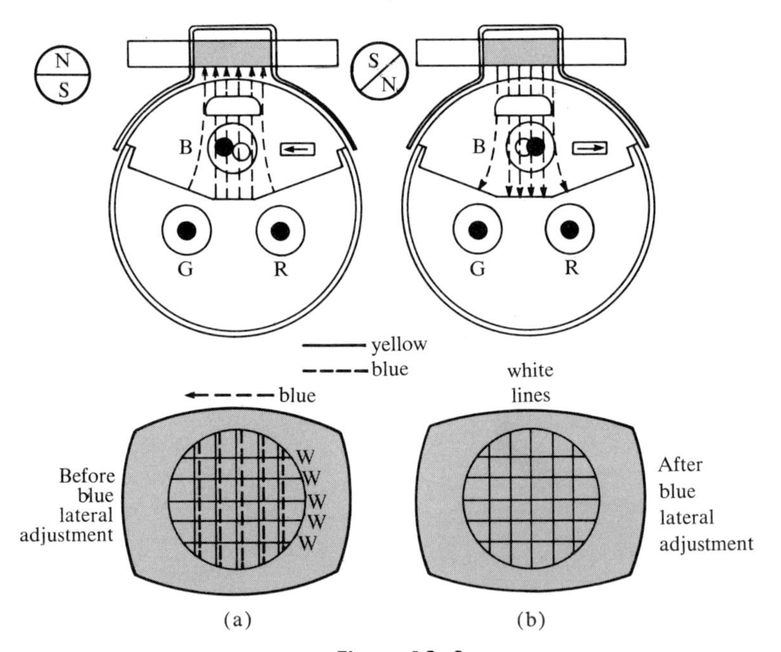

Figure 10–3

netic strength and will, in some cases, minimize purity change with the adjustment of the blue-lateral magnet.

We have now provided the necessary external adjustments on the three beams of the tube to make it do what it was originally designed to do. The purity magnet (chapter 9) adjusts the proper approach angle of all three beams to the shadow mask, so they will each see the proper color phosphor. The convergence magnets position each of the three beams individually for a radial position in the tube, and the blue-lateral magnet gives the blue beam an additional correction in a lateral direction. With these adjustments, all three beams will hit the same spot at the center of the screen at the same instant. All of these adjustments are known as *static adjustments* and are made while viewing the center of the screen.

Dynamic-Convergence Corrections

Other corrections have to be made as the three beams sweep to the outer edges of the screen. These are known as *dynamic-convergence corrections.*

Figure 10–4 shows that all three beams are converged at the center. Not only are the triads not converged both vertically (top-to-bottom) and horizontally (side-to-side) but they do not form equilateral triangles. The yoke windings are distributed to produce a field which is

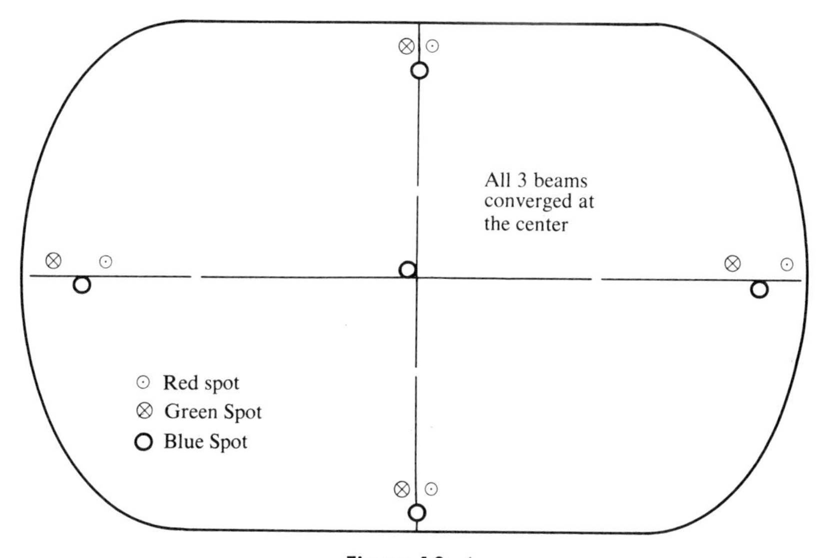

Figure 10–4

stronger at the edges than at the center. This is so designed to overcome the length of deflection due to the geometry of the red and green guns. The result of this uneven distribution of the yoke field will be to bring red and green closer together as is required.

Figure 10–5 shows the result on the overall rasters when only the center is statically converged. Each color gun sweeps its own raster as if it were the only one in existence. In other words, the three gun deflections act as though they were three separate color tubes. Dynamic convergence corrects for the outer extremities of misconvergence.

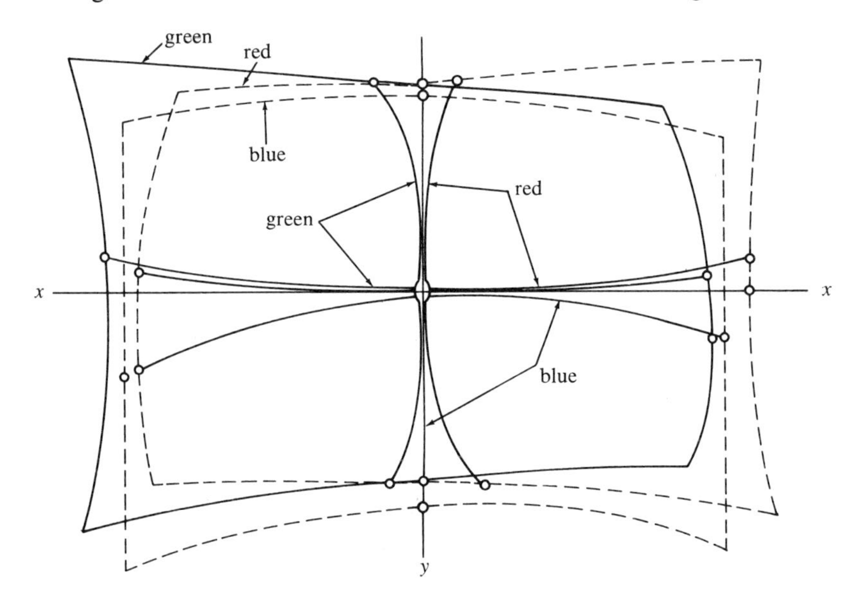

Figure 10–5

Misconvergence Due to CRT Geometry

Figure 10–6 describes with more detail the misconvergence caused by the picture-tube geometry. The three rasters are made to converge at the center of the screen through adjustment of the static magnets. Since the three rasters are projected on a nearly flat faceplate, the distance from the yoke deflection center to the outer edges of the screen is greater than it is to the center. When the deflection system moves the three beams to the edge of the screen, the beams converge at a point before reaching the mask. When the three beams reach the mask, they have crossed and light the screen in three different places. Figure 10–6 shows that the misconvergence caused by the tube is greatest at the edges of the screen. This can be corrected by making the focal length of the three beams variable, in step with the vertical-sweep system and horizontal-sweep system.

Coils wound on the convergence-magnet assemblies are placed over each electron gun. The passing of a current through the coils causes the focal length, or the point where the beams cross over, to change. The amount of current required for convergence is greatest when the sweep is at the edges of the screen and gradually reduces until it becomes zero at the center where correction is not needed. Two coils are used on each convergence-magnet assembly. One is used to correct misconvergence in a vertical plane, and the other is used in the horizontal plane.

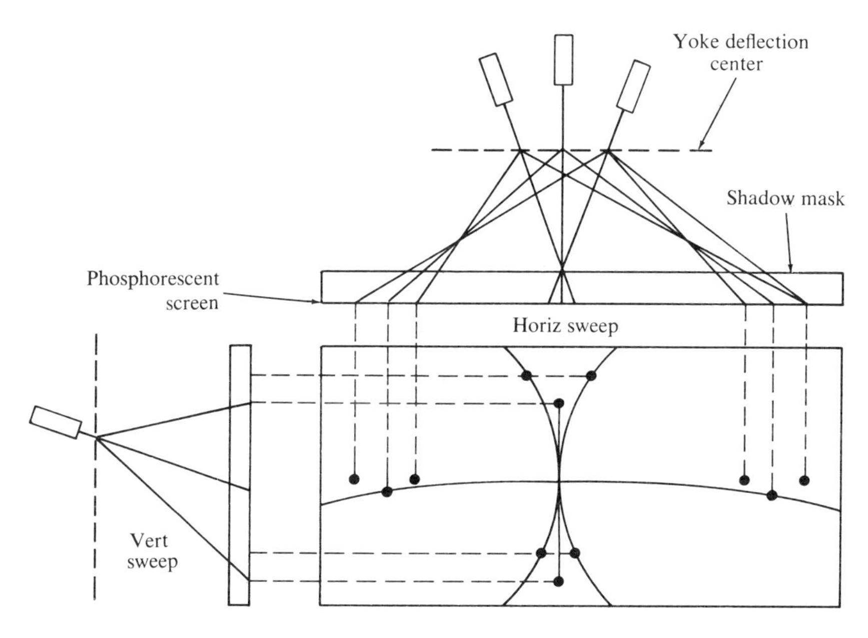

Figure 10–6

Figure 10–7 shows the corrective wave form of current and its effect on focal length. A 60-Hz and 15,750-Hz current of the wave form shown in Figure 10–7 is provided to maintain convergence as the tube is scanned vertically and horizontally. The 60-Hz parabola corrects vertical-convergence errors, and the 15,750-Hz parabola corrects horizontal-convergence errors. The amount and the wave shape of the current required for convergence will vary slightly from set to set, because of slight differences in the picture tube and other variables. Two controls are provided for each parabolic correction current, so that they can be adjusted for the particular picture-tube and yoke combination. One control, called an *amplitude control,* adjusts the amplitude or the amount of current. The other control, called a *tilt control,* tilts the wave form of the current to insure that all points are converged. For convenience, red and green convergence coils are usually connected in series. This necessitates the addition of balance controls called *differential-tilt* and *differential-amplitude controls.*

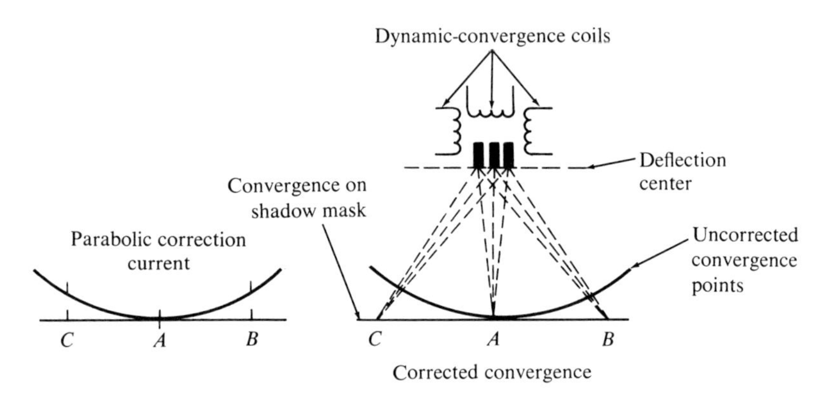

Figure 10–7

Figure 10–8(a) shows the action of an amplitude control. This control adjusts the magnitude, or amount, of correction current through each convergence coil. The action of a tilt control is shown in Figure 10–8(b). The tilt circuit adds a sawtooth current to the parabolic current and alters wave shape of the current as shown. The tilt control adjusts the amount and the polarity of the sawtooth current and becomes a waveform adjustment.

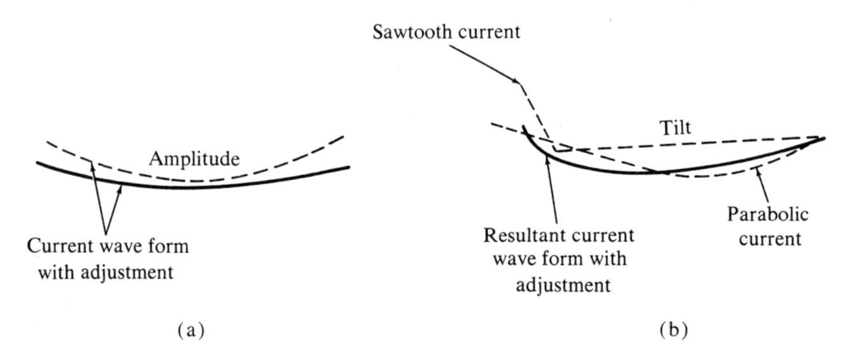

Figure 10–8

Vertical Correction

The voltage for vertical dynamic convergence is obtained from a sawtooth voltage which is present on the cathode of the vertical-output tube. Figure 10–9 shows the basic circuit for producing the parabolic convergence current.

The vertical-sweep sawtooth current that passes through the vertical-output tube will cause a sawtooth voltage to appear across the unbypassed cathode resistor. This sawtooth voltage passes through a coupling capacitor (*C*) and is applied across the convergence coil. If the convergence coils were a pure inductance, this sawtooth voltage would produce the required parabolic current. The convergence coils, however, are resistive as well as inductive, so the sawtooth voltage must be modified in order to produce a parabolic current through the coil. Diode E-1 limits the negative peak voltage as shown in Figure 10–9. This produces a wave form of the correct voltage to cause a parabolic current through the convergence coil. The amplitude control determines the voltage across the coil and, consequently, the amount of current.

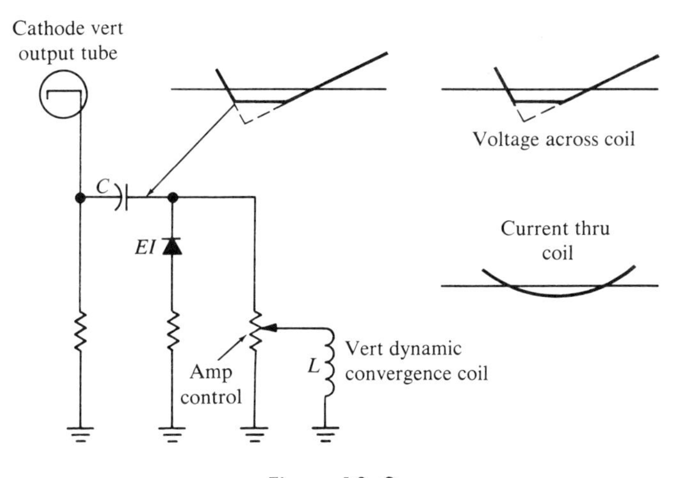

Figure 10–9

The red and green electron guns occupy similar positions in the yoke deflection field, and their dynamic-convergence circuits may be combined and fed from a single voltage source. Figure 10–10 shows the vertical red-green dynamic-convergence circuit and the beam movement.

With the red and green dynamic-convergence coils connected in series, adjustment of the amplitude control will cause an increase or decrease of current in both coils.

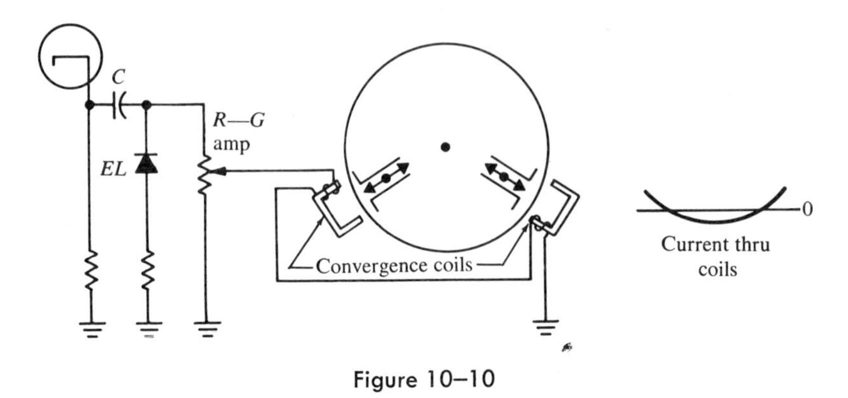

Figure 10–10

Figure 10–11 shows the effect viewed on the screen when a parabolic current flows through the dynamic-convergence coils.

In Figure 10–11(a) are shown vertical red-green (R-G) lines passing through the center of the screen before the dynamic correction is applied. The center of the line (point *C*) has been converged using the static adjustments. Without any dynamic correction, the misconvergence becomes progressively worse toward the edges of the screen (as indicated). If we pass a current of the wave form indicated through the convergence coils, the beams will move in the direction indicated by the arrows. Point *A* on the current curve has maximum current and causes maximum displacement of the beams at point *A* on the scan line.

At point *B* (current curve) the current is zero, so there is no action at point *B* on the scan line. At point *C* the current has reversed, causing the center of the screen to misconverge in the direction indicated by the arrows at the center of the scan line. Again, at point *D* the current passes through zero, so there is no movement at point *D* on the scan line. The current becomes maximum at point *E* and causes beam movement as indicated by the arrows at point *E*. It should be noted that the beam movement at each edge of the screen (toward convergence) is greater than the movement (away from convergence) at the center of the screen. The diode in the wave-forming circuit limits the peak voltage and, consequently, the current at the center of the screen. This allows a maximum of convergence action at the edges of the screen and a minimum of deconvergence at the center.

Once the amplitude is correctly adjusted, red should be either converged on green along the vertical line or symmetrically displaced as indicated in Figure 10–11. If parallel lines are obtained, they may be easily converged with the static adjustments.

The red and green vertical dynamic coils are connected in series and return to ground through a control connected across a secondary winding on the vertical-output transformer. The deflection current flowing in the transformer will produce a pulse of voltage across the secondary winding. The wave form across the winding is quite different from the sawtooth voltage developed across the cathode resistor of the vertical-output stage. The inductive discharge from the vertical-deflection yoke causes a pulse of voltage to appear across the secondary winding during vertical retrace. The voltage pulse does not appear across the cathode resistor, since the tube is not conducting (biased beyond cutoff) during vertical retrace.

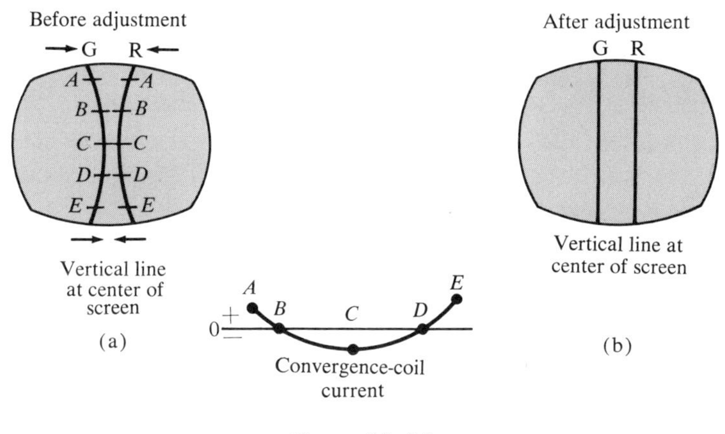

Figure 10–11

The wave form of the secondary voltage is shown in Figure 10–12. At opposite ends of the winding, equal voltages of opposite polarity appear. With the connection of a control across the winding, varying amounts of positive or negative pulse can be selected.

The pulse of voltage across the convergence coils causes a sawtooth current to flow. When the sawtooth current is added to the parabolic current, wave forms as shown in Figure 10–12 are produced. The pulse of voltage occurs during vertical retrace, so the greatest change in current wave form will be at the beginning of the vertical retrace or at the top of the screen.

The red-green vertical line through the center of the screen is converged from the center down, but is either underconverged or overconverged from the center up. It is apparent that the dynamic-current wave form must be altered to produce the desired correction at the top of the screen without deconverging the bottom.

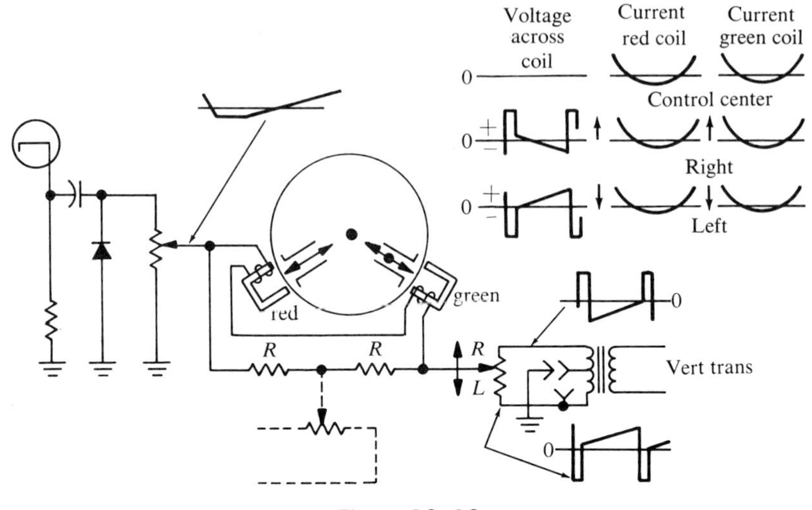

Figure 10–12

In Figure 10–13(a) red and green are underconverged from the center up. The tilt control is adjusted to add a positive pulse across the convergence coils. The positive pulse will increase the current through the convergence coils at the beginning of the vertical trace and will cause the red and green beams to move toward convergence as indicated by the arrows.

In Figure 10–13(b), red and green are overconverged. The tilt control is adjusted to produce a negative pulse across the coils which will reduce the current at the beginning of the vertical trace. The reduction in current will allow the beams to move toward convergence as indicated in Figure 10–13(b).

Since the red and green convergence coils are connected in series, a balance control is provided to insure that the current in each coil is equal. This control will allow for variations in the coils and the cathode-ray tube (CRT). The circuit is shown in Figure 10–14. The balance, or differential, amplitude control is connected so that a portion of the resistance is across each coil. As the control is adjusted, resistance is taken from across one coil. It is placed across the other. Since the resistance across the coils is adjustable, the distribution of current through the coils can be controlled.

As the current through the red and green convergence coils is increased with the amplitude control, a condition seen in Figure 10–15(a) may be encountered. In this case, the current in the two coils is not equal

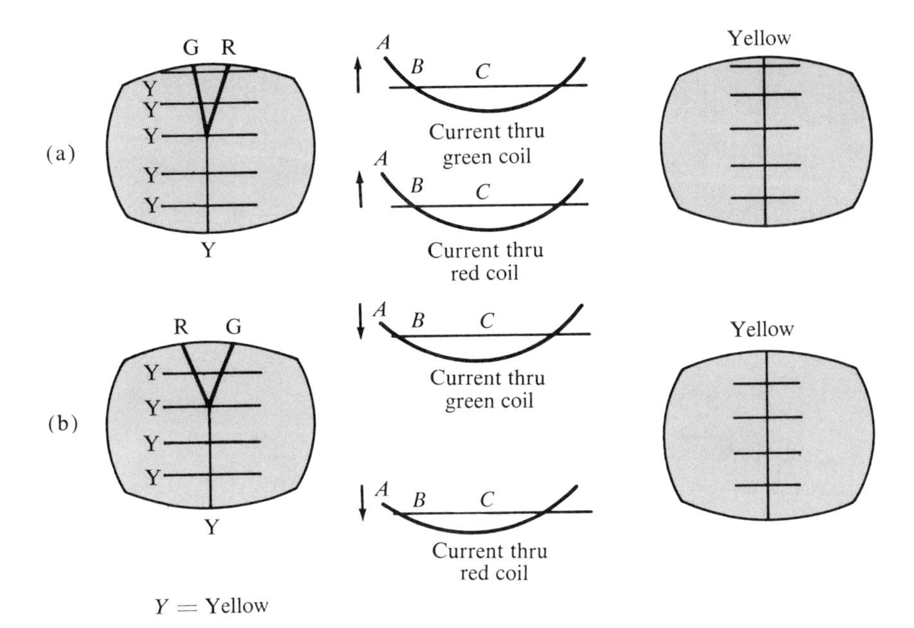

Figure 10–13

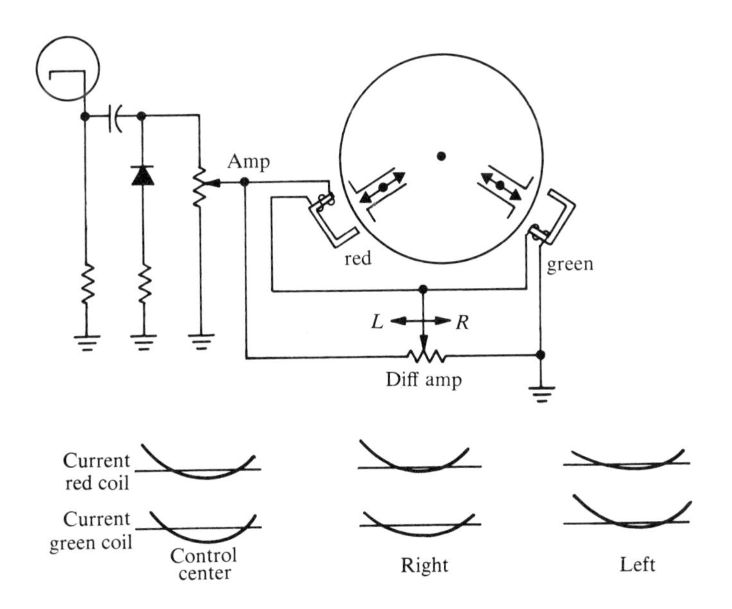

Figure 10–14

and green is moved further than red. With this condition, the beams are converged along a vertical line as in Figure 10–15(b). However, examination of horizontal lines near the top-center and bottom-center of the screen will show that red and green are vertically displaced. Also, the vertical line has a slight bow in it. If the differential-amplitude control is adjusted so that the current through the green coil is decreased, the beam movement will be as indicated in Figure 10–15(b). This control is adjusted for convergence of red and green along horizontal lines near the bottom-center of the screen. Adjustment is made for either convergence or symmetrical displacement as in Figure 10–15(c) which can be converged with static adjustments.

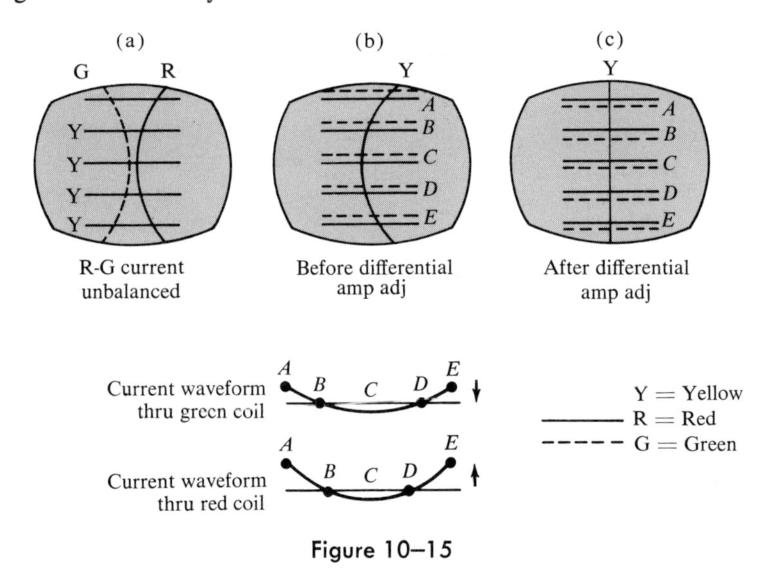

Figure 10–15

The fourth, and final, circuit needed to provide vertical red-green convergence is a differential-tilt circuit. The basic circuit is shown in Figure 10–16.

The differential-tilt control is a balance control for current wave form. The previously discussed tilt circuit adds a pulse of voltage across the red-green convergence coils, and the differential-tilt control adjusts the tilt-voltage division between the coils.

To accomplish this balancing action, the differential-tilt circuit provides opposite tilt action to current wave forms through the red-green dynamic-convergence coils. A center-tapped secondary winding on the vertical-output transformer supplies varying amounts of either a positive or a negative pulse voltage, depending on the setting of the differential-tilt control.

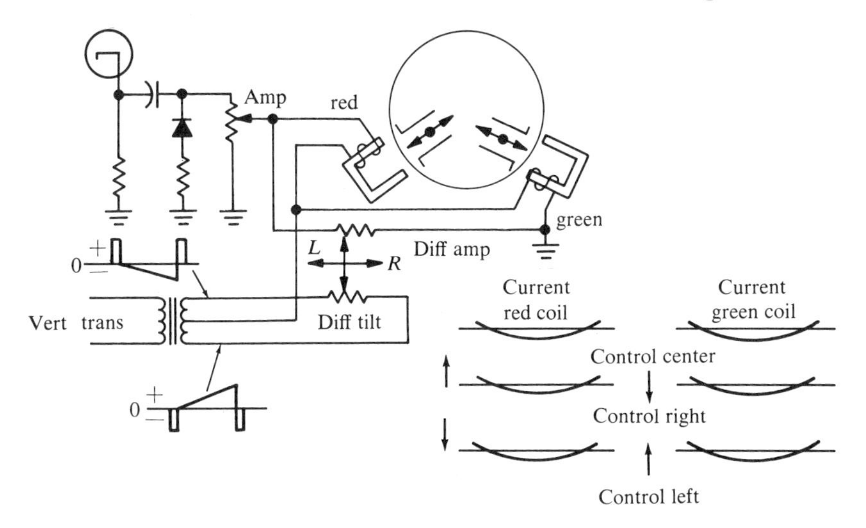

Figure 10–16

Figure 10–17 shows an equivalent red-green differential-tilt circuit. A center-tapped secondary winding on the vertical-output transformer is connected as shown. A positive pulse will appear from the center tap to one end of the secondary winding and a negative pulse from the tap to the other end of the winding.

If the differential control is adjusted to the left as shown, a negative-going pulse will appear at the arm of the control, (the center tap is the

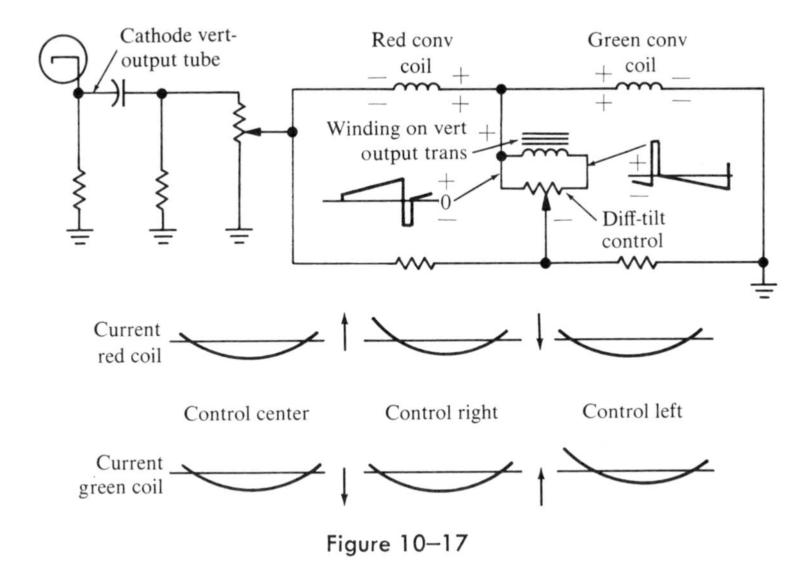

Figure 10–17

reference point). This pulse causes a voltage to develop across the two convergence coils as indicated by polarity signs above the two coils. The sawtooth voltage from the cathode of the vertical-output tube also causes a voltage to develop across the coils, as indicated by the polarity signs below the coils. When these voltages are combined, the current through the red coil is reduced and the current through the green coil is increased. Since the voltage supplied by the winding on the output transformer is of a pulsed nature and occurs during vertical retrace, the major current changes occur at the beginning of the vertical trace or at the top of the screen. With the control to the left, the wave form of the current will be as indicated in Figure 10–17. When the control is moved to the right, opposite action will occur, and the tilt is in the opposite direction.

To make the vertical-differential adjustment, a horizontal line near the top-center of the screen is observed. Figure 10–18(a) shows red and green converged along a vertical line through the center of the screen; however, red and green are displaced along a horizontal line in the top-center of the screen. Red is underconverged. The differential-tilt control is adjusted to increase red coil current, and reduce green coil current at the beginning of the trace. The two beams will move toward convergence as indicated by the arrows.

Figure 10–18(b) shows the opposite condition. The control is adjusted in the opposite direction and beam travel is in the direction shown.

The four controls needed for vertical red-green dynamic convergence are amplitude, tilt, differential amplitude, and differential tilt. The amplitude control adjusts the amount of current through the coils. The tilt control adjusts the wave form of the current. The differential controls are balance controls made necessary because the two coils are connected in series. They balance the current and wave form between the two coils.

The blue electron beam also requires a vertical-dynamic current in order to converge it on red-green, now in convergence. The blue electron beam occupies a position in the yoke field which is unlike either of the other two beams, so it must be controlled by separate circuits. Since there is only one convergence coil, balance or differential controls are unnecessary. Only an amplitude and a tilt control are needed. The basic circuit is shown in Figure 10–19.

The voltage developed at the cathode of the vertical-output tube is also used to supply the blue vertical-dynamic circuit. A control is connected in series with the red-green amplitude control. A sawtooth voltage of variable amplitude is available at the arm of the control. The diode modifies this voltage so that it will produce a parabolic current through the blue convergence coil. The coil returns to ground through a center-tapped secondary winding on the vertical-output transformer. This wind-

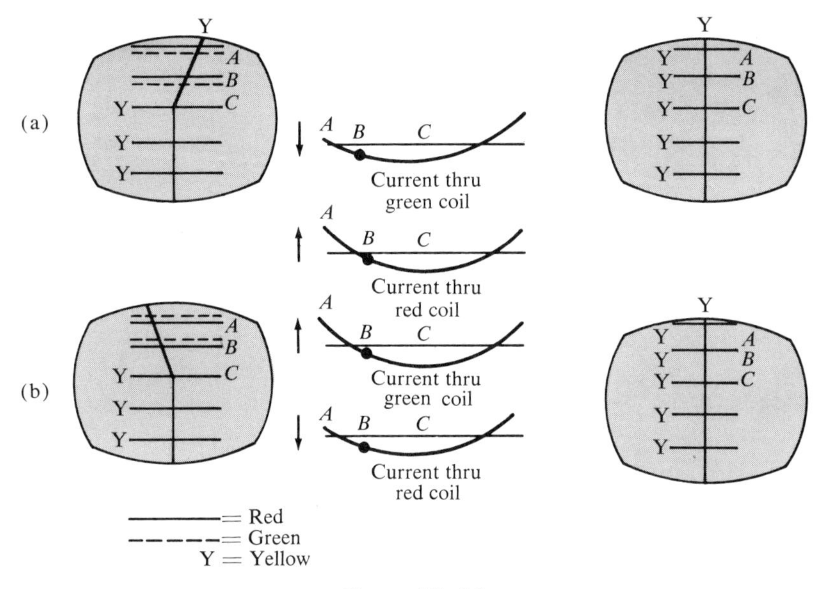

Figure 10–18

ing supplies a positive or negative pulse as selected by the tilt control and applies it across the blue convergence coil. This voltage adds to the sawtooth voltage across the coil and provides an adjustment of the wave form of current in the same manner as the red-green tilt adjustment.

Figure 10–20 shows the effect of increasing the blue dynamic amplitude. A blue dynamic-amplitude adjustment is made by observing hori-

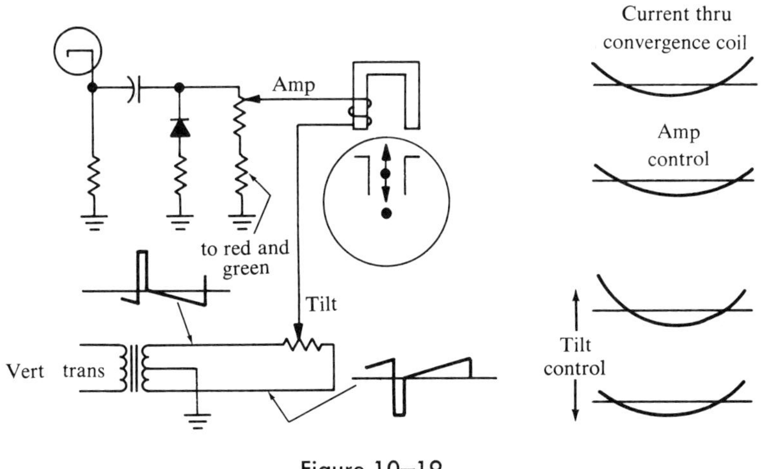

Figure 10–19

zontal lines near the top- and bottom-center of the screen. Figure 10–20 shows the normal misconvergence of blue before any dynamic correction is made. Blue is converged on red-green at the center of the screen using the static adjustments. Before dynamics are added, blue will be displaced from red-green at the top- and bottom-center as shown. As the amplitude control is advanced, the sawtooth voltage will cause a parabolic current to flow through the blue convergence coil as shown. Blue will be displaced upward at the top and bottom of the screen and will be displaced downward at the center of the screen. The amplitude control should be adjusted for either convergence of blue on red-green along horizontal lines through the center of the screen, or for symmetrical displacement as shown in Figure 10–20. Once blue is symmetrically displaced from red-green, it can be converged using the static adjustments.

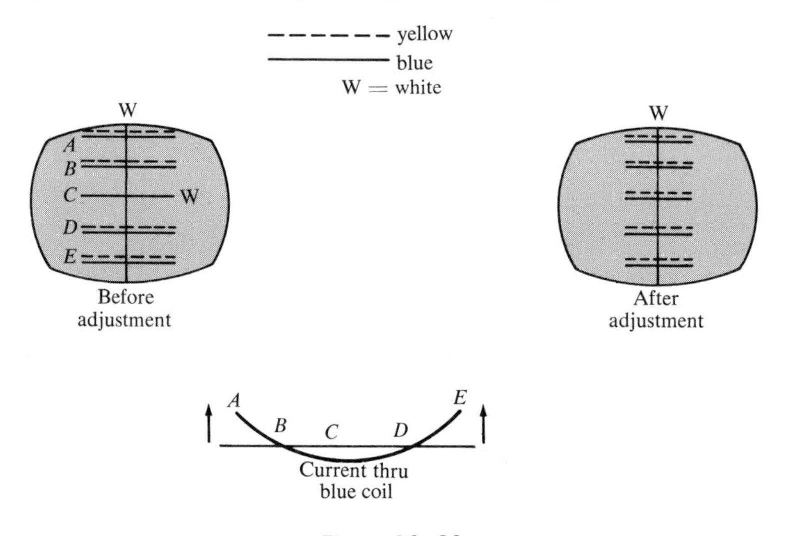

Figure 10–20

If, after blue amplitude is adjusted, it is not possible to converge blue on red-green, it may be necessary to change the wave form of the blue convergence current by adjusting the tilt control. Figure 10–21 demonstrates the need for tilt adjustment.

Blue is converged from the center down but is displaced in the top-center of the screen. The tilt control is adjusted so that current through the convergence coil is increased at the beginning of the trace. Blue will be displaced upward in the top-center of the screen where the misconvergence was present.

The complete vertical dynamic-convergence circuit is shown in Figure 10–22.

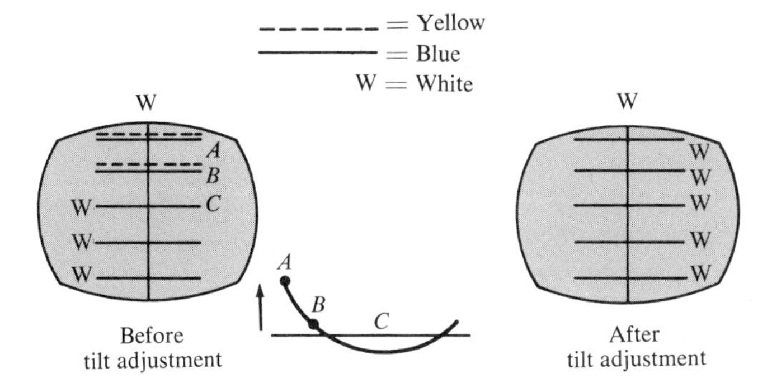

Figure 10–21

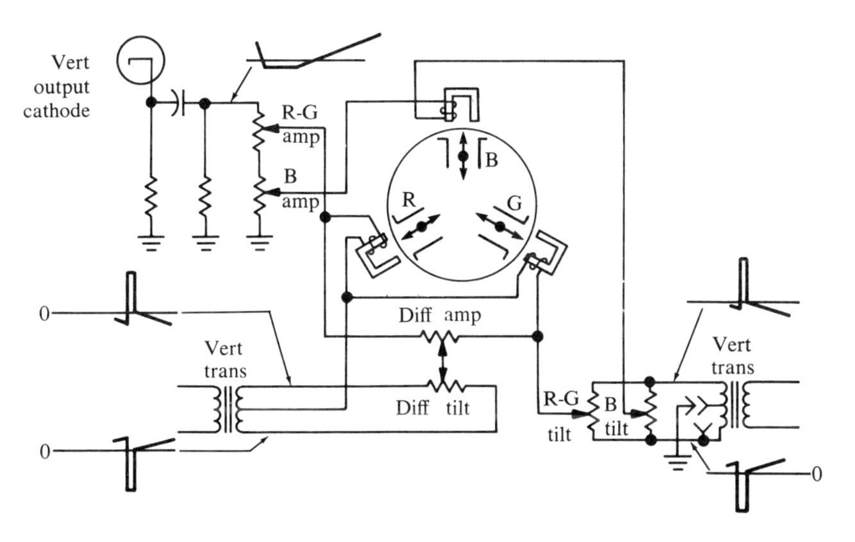

Figure 10–22

Horizontal Correction

The previously discussed convergence circuitry serves to maintain convergence as the screen is scanned vertically. A dynamic correction (synchronized with the horizontal-sweep system) is necessary to maintain convergence with horizontal sweep. To accomplish this, circuitry is provided to produce a parabolic current that is locked in time with the horizontal-scan system. Controls are provided to adjust the amount (or amplitude) of current and the wave form. The action of the horizontal dynamic-convergence controls is quite similar to the equivalent vertical control. The area of the screen to be viewed during adjustment, however, is different.

The circuitry to produce a parabolic current at the horizontal-scan frequency is different from that used to produce a parabolic current at the vertical frequency for two reasons. The higher horizontal-scan frequency (approximately 15 kHz) makes it practical to use resonant circuits for increased efficiency. The increased efficiency will reduce the amount of power taken from the horizontal-sweep system. The second difference is the wave form of the voltage available in the horizontal deflection system. The vertical system contains a sawtooth voltage which is near the required wave form. The voltage available in the horizontal system is a pulse of short duration caused by the inductive discharge of the deflection yoke. Circuitry is provided to change the pulse to the proper wave form to cause a parabolic current in the convergence coils.

Figure 10–23 shows the basic circuit for blue horizontal dynamic convergence. Inductance L_1 and the impedance of the convergence coil forms the pulse from the horizontal-output transformer into a sawtooth voltage. This sawtooth voltage appears across the convergence coil. Inductance L_1 is made adjustable so that the voltage across the convergence coil can be varied. The current through the coil will vary with the voltage and provide current-amplitude control.

The convergence coil is made parallel resonant by capacitor C_1. This increases the efficiency of the convergence circuit and reduces the amount of power taken from the horizontal-scan system. The coil must be resonant at a frequency somewhat lower than the scan frequency. It is resonated so that it completes one-half cycle in the time it takes the beam to scan from the left to the right side of the picture tube.

Figure 10–23 shows the pulses from the horizontal transformer, the developed sawtooth voltage, and the resulting current.

A control R_1 is provided to adjust the shape of the wave form of the current through the convergence coil. This is accomplished by connecting a variable resistor in the tuned circuit formed by C_1 and the convergence coil. As the control is adjusted to position *A* (Figure 10–23), resistance is removed from the circuit and when adjusted to position *B* will add maximum resistance to the circuit. As resistance is added, the current wave form through the coil will become more like the wave form of the applied voltage which is saw shaped. As it becomes more saw shaped, the current at the beginning of trace (left side of screen) becomes greater than at the end of trace. With the control in the other direction, the wave form of the current becomes less sawlike and the current amplitude at the beginning of trace is less than at the end of trace. (The wave forms of current that result are shown in Figure 10–23.) Since this control changes the wave form of current at the beginning of trace, its effect on convergence is viewed at the left side of the screen.

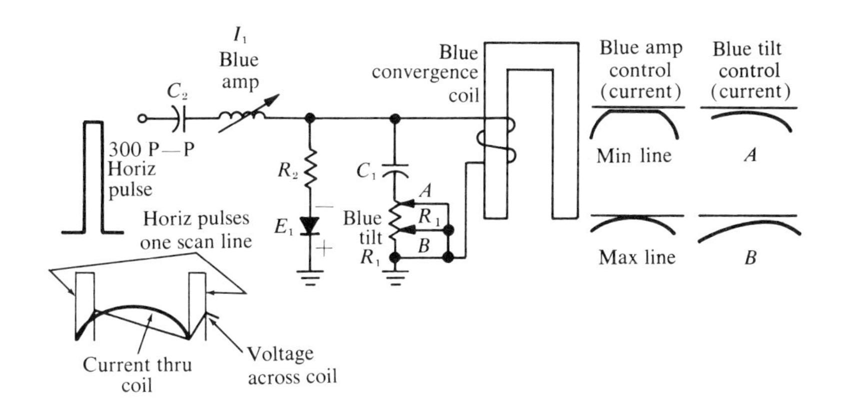

Figure 10–23

Each of the horizontal-dynamic circuits employs a system of dc clamping to maintain center convergence, as the edges of the screen are brought into convergence. This adds to the ease of adjustment and eliminates the necessity of making static adjustments as the horizontal dynamics are adjusted. Figure 10–24 shows an equivalent dc clamping circuit.

As established earlier, the pulse from the horizontal transformer is integrated into a sawtooth-shaped voltage by *L* and *R* (see Figure 10–24). Without diode *E,* the ac axis (the electrical center of the developed

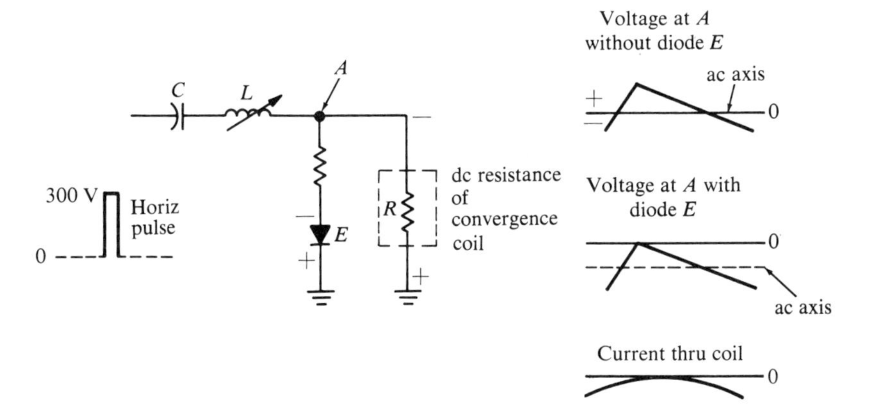

Figure 10–24

sawtooth voltage) would be zero after passing through capacitor *C*. That is, a portion of the sawtooth would be positive and the other portion negative.

With the diode connected, the positive portion of the sawtooth voltage will cause the diode to conduct and produce a negative dc voltage at point *A*, or across the resistance of the convergence coil.

The negative voltage developed by the diode will displace the ac axis downward as shown so that the peak of the sawtooth voltage is zero and all other portions are negative. If the amplitude control is adjusted to increase the amplitude of the sawtooth voltage, a greater negative voltage will develop at point *A*, moving the ac axis still further negative. In this manner, the peak sawtooth voltage is clamped at the zero line. The parabolic current that flows as a result of the negative sawtooth voltage will also have its peak clamped at the zero line as shown. The peak of the parabola occurs at the center of the sweep or the center of the screen. Because of the clamping action, the convergence current is always zero at this point, so convergence at the center of the screen is not affected by horizontal-dynamic adjustments.

Figure 10–25 shows blue deconverged from red-green on a horizontal line through the center of the screen, as it normally is after static center convergence, but before dynamic-convergence corrections are added. As the blue horizontal amplitude is adjusted to increase the current through the convergence coil, blue is deflected upward as shown by the arrows. Maximum change occurs at points *A* and *B* since current through the coil is greatest at these two points. The dc clamping makes the current travel through the coil zero at point *B*, so center convergence is not affected.

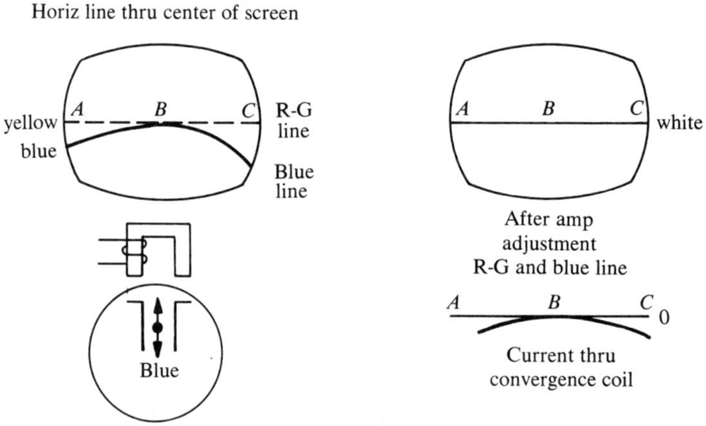

Horiz line thru center of screen

yellow A B C R-G line

blue

Blue line

Blue

A B C white

After amp adjustment R-G and blue line

A B C 0

Current thru convergence coil

Figure 10–25

The amplitude control is adjusted so that blue is converged at all points between *B* and *C* on the horizontal line.

After blue is converged with red-green on the right side of the screen, with the blue amplitude adjustment, the left side may not be converged.

Two conditions are shown in Figure 10–26. Blue is underconverged in 12–26(a) and overconverged in 12–26(b). The blue horizontal-tilt control will either increase or decrease the current through the convergence coil at the beginning of the trace, so will correct misconvergence on the left side of the screen. The left side of a horizontal line through the center of the screen is observed and the tilt control is adjusted for convergence of blue on red-green.

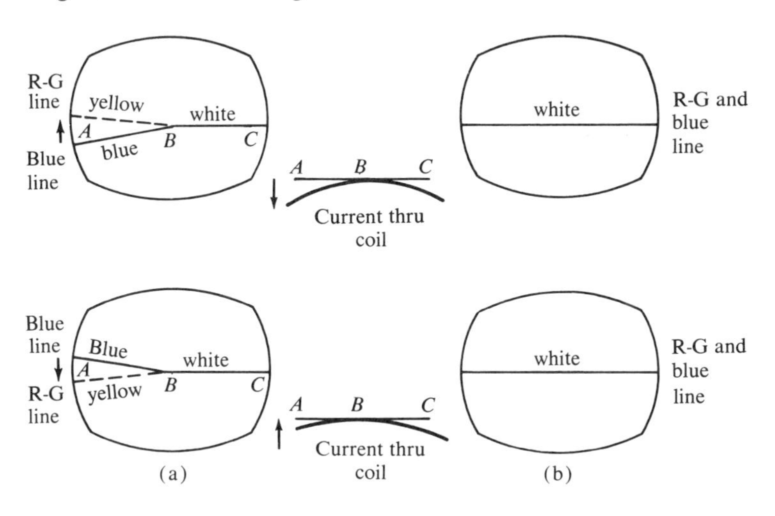

Figure 10–26

The circuitry required for red-green dynamic convergence is quite similar to the previously discussed blue horizontal-convergence circuit. Since the red and green electron beams occupy the same relative position in the yoke field, they require the same type of convergence correction and may be fed from a common voltage source. Figure 10–27 shows the dynamic red-green convergence circuit.

The red-green horizontal dynamic-convergence circuitry consists of two circuits of the type used for blue, connected in parallel and fed from a single voltage source. Each convergence coil is made parallel resonant at 12 kHz by capacitor *C*. Operation of the amplitude and tilt circuitry is exactly the same as previously discussed in the blue horizontal-dynamic circuit. Inductance L_1 is adjustable and controls the amount of voltage across the two convergence coils and consequently, the amplitude of current. Control R_2 adds or removes resistance from the two tuned circuits simultaneously to provide the necessary tilt action. As was the case with red-green vertical-convergence circuits, two circuits are grouped to provide controls that balance the current and wave forms between the two coils. These two controls are called *differential amplitude* and *differential tilt.*

The adjustable inductive voltage divider, L_2, determines how the voltage divides between the two convergence coils. The balance of current between the coils can be varied by adjusting the voltage division.

The differential-tilt control operates in the same manner as the tilt control except that as resistance is added to one tuned circuit, it is removed from the other and causes opposite tilt action to the two convergence-current wave forms. Both the tilt and the differential-tilt controls vary the wave form of current at the beginning of trace or at the left side of the screen.

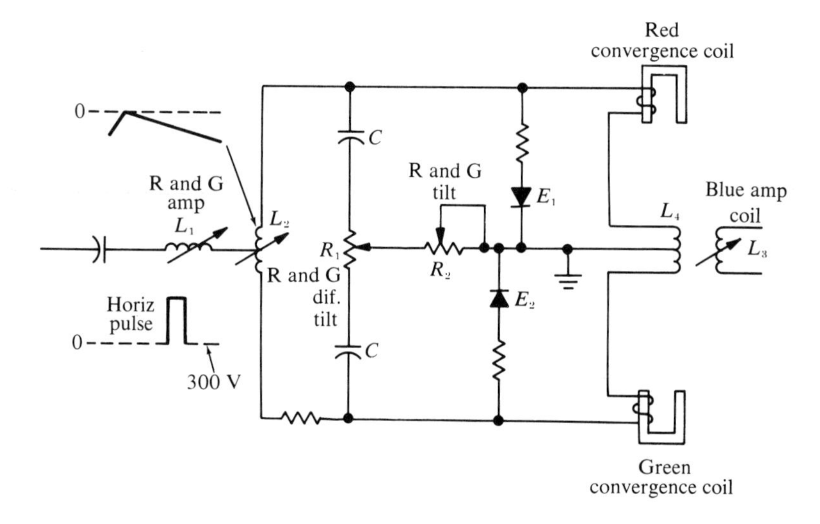

Figure 10–27

A red-green horizontal line through the center of the screen is shown in Figure 10–28(a). Red and green are converged along a horizontal line, but examination of vertical lines at either end of the horizontal line will show that they are misconverged in that plane. This is the normal misconvergence of red and green before any dynamic corrections are applied. As the horizontal red-green amplitude control is adjusted to increase the current through both convergence coils, the electron beams will move in the direction indicated by the arrows.

Figure 10–28(b) shows the current through the red-green coils. The current is maximum at points *A* and *C* where maximum correction is needed. The dc clamping causes the current to be zero at point *B*, so convergence at the center of the screen is not affected. The amplitude control is adjusted for convergence of red on green while viewing vertical lines on the right side of the screen.

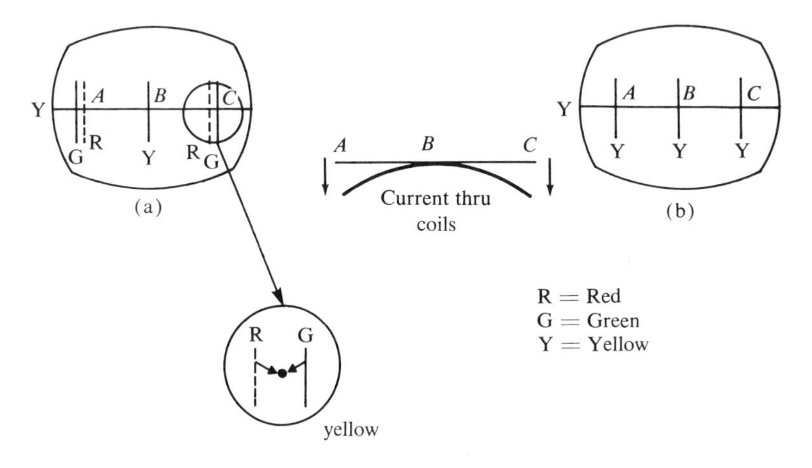

Figure 10–28

After the red-green amplitude is adjusted for convergence of vertical lines on the right side, there is a possibility that vertical lines on the left side of the screen may not be converged. An example is shown in Figure 10–29. To correct this condition, an increase in current through the

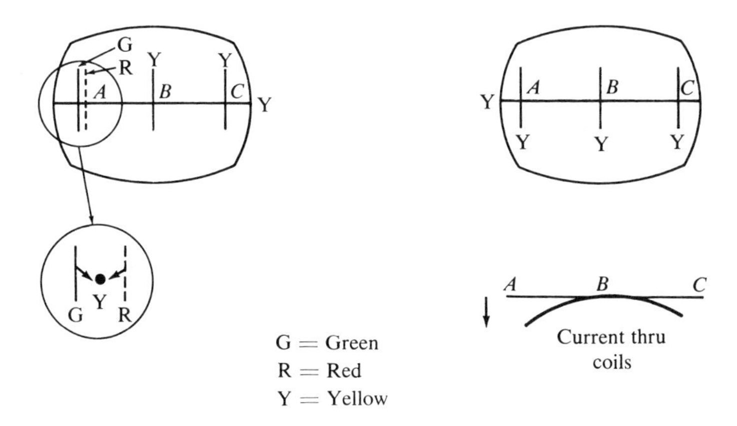

Figure 10–29

convergence coils is required at the beginning of the horizontal trace. The tilt control provides this action. As the control is rotated toward maximum resistance, the current at point *A* (current curve, Figure 10–29) will increase while there will be little or no change of current at point *C*. This will allow convergence on the left side without deconverging the right side of the screen. The dc clamping circuit will maintain center convergence.

After red is converged on green through observation of vertical lines on either side of the screen and adjustment of horizontal amplitude and tilt, they may not be converged along a horizontal line. An example is shown in Figure 10–30. The deconvergence along the horizontal line is a result of an imbalance of current between the red and green convergence coils. The differential-amplitude control is used to adjust or balance the two currents.

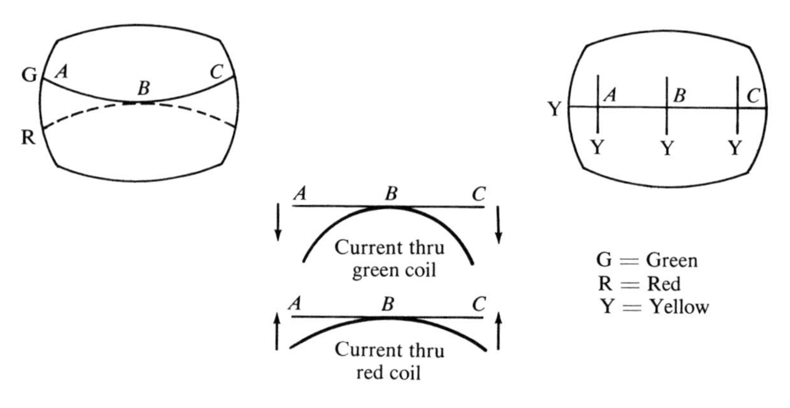

Figure 10–30

In the above example, the green coil needs additional current and the red coil needs less current in order to bring red-green into convergence. The differential-amplitude control is adjusted for convergence by viewing horizontal lines on the right side of the screen. The convergence action is as indicated by the arrows.

After the differential-amplitude control is adjusted for convergence of red-green on horizontal lines on the right side of the screen, there may be some convergence error on the left side. An example is shown in Figure 10–31. This condition is caused by a wave-form imbalance between the two coils. The differential adjustment will redistribute the tilt voltage between the coils and increase the tilt action in one and decrease it in the other.

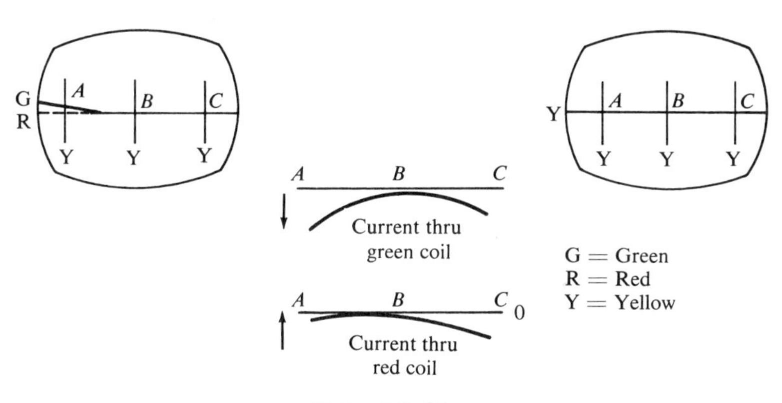

Figure 10–31

Correction of the misconvergence, in the example shown, will require an increase of current in the green coil and a decrease in the red coil at the beginning of the trace. This adjustment is made while viewing red-green horizontal lines on the left side of the screen. Beam movement is in the direction indicated by the arrows.

From the preceding discussion of the dynamic convergence circuits, it should be apparent that the amplitude controls will affect convergence in two areas of the screen while the tilt controls will affect only one area. If the dynamic adjustments are made in the proper order, the set may be converged by viewing only one area of the screen at a time.

questions

1. The ceramic disc permanent magnet converges the raster at the center. True _____ False _____
2. In what direction does the blue magnet move the blue lines?
3. In what direction does the red magnet move the red dot?
4. In what direction does the green magnet move the green dot?
5. The blue-lateral magnet moves the blue dot in what direction?
6. The movement of the dots by the disc magnets is called (static), (dynamic) convergence.
7. Explain why dynamic convergence is necessary.
8. Since the convergence errors are parabolic in shape, what type of correction shape is necessary for a straight line?
9. For the correction of what type of convergence error is the tilting of the parabolic wave shape necessary?
10. When the best that can be accomplished with dynamic adjustments are straight parallel lines, the disc magnets should then be used to converge these lines into one. True _____ False _____

part two

FUNDAMENTALS OF SERVICE

11 basic transistor theory

Television manufacturing has progressed from use of vacuum tubes to use of solid-state (transistor) devices. Another important development has been the new plug-in panel. Motorola, with their Quasar* and Works-in-a-drawer,* was the first to develop a domestic television receiver employing a plug-in panel. Most manufacturers are now using this concept which permits the customer to keep his color receiver at home during repairs. Only the "drawer" need visit the serviceman's bench. With this trend toward the solid-state chassis, a radical change in the home service technique has taken place.

Although Motorola's model TS 915 is the prototype used in the following chapters, the basic solid-state service techniques hold true for all receivers using these new concepts.

---

* Trademarks of Motorola, Inc.

Atoms, Electrons, and Holes

Scientists believe that the basic building block of all matter is the atom. The individual atom is similar to a miniature solar system in which small particles orbit a main body, much like the planets orbit the sun in our universe. The orbiting, negatively charged particles are called electrons. The body around which they orbit is called the nucleus (Figure 11–1) There are small, positively charged particles in the nucleus called protons. An equal number of electrons and protons exactly balance, or neutralize, the electrical charge. For example, the element silicon has 14 protons in the nucleus, balanced by 14 electrons in orbit.

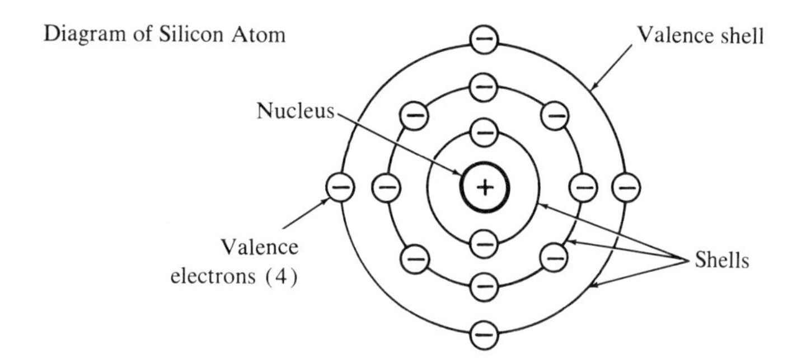

Figure 11–1

An orderly, regularly repeating pattern of atoms forms a *crystal*. Practically all matter is crystalline—some less so than others. Clay, for example, is a poorly crystallized substance. Iron, is highly crystallized. The difference between clay and iron is caused by the difference in their atom arrangements. In clay the atoms are disorganized. In iron the atoms are highly organized and tightly bonded together.

The orbiting paths taken by the various electrons are called shells. It is the outer shell that is important to our study of transistors. The outer shell is called the valence shell. Electrons in the valence shell are called valence electrons. These valence electrons bind the atoms together in the crystal, forming covalent bonds between adjacent atoms in the structure. The silicon atom has four valence electrons. In pure silicon covalent bonding exists because all the atoms in the crystal have four valence electrons.

Pure silicon is an insulator because there are no free, or "extra," electrons in the crystal to create a current. In vacuum tubes, a supply of free

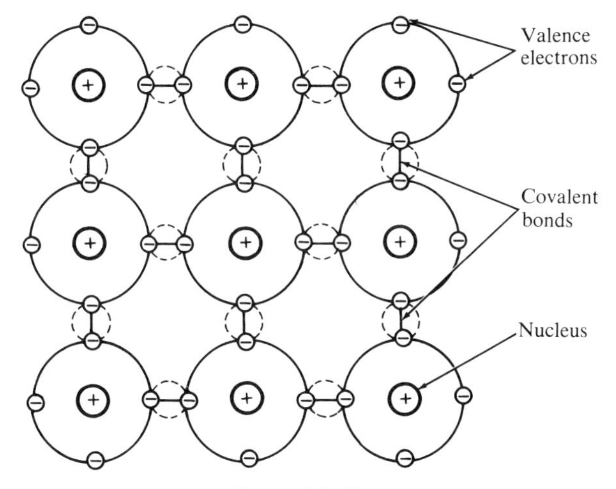

Figure 11–2

electrons is made available through application of heat to a filament, or cathode, which boils electrons out of the valence shells.

To make free electrons available in silicon, a different technique is used. An element having a different number of valence electrons is added to pure silicon in a controlled, small quantity. The additional element, called an impurity, represents only one part per ten million and does not change the crystal structure. If the impurity added to the silicon has five valence electrons, covalent bonds cannot be formed in all the silicon atoms (which have four such valence electrons). This causes a game of musical chairs in which one electron gets left "without a chair," or bond, and is available to provide electron current. The basic silicon is no longer an insulator, but is now a conductor. Because it is not a good conductor, the silicon is called a *semiconductor*. Arsenic and antimony have five valence electrons and are typical donor additives.

The silicon is called *N-type* because it has an excess of negative electrons which are not bound in the crystal lattice and are free to move (Figure 11–3).

Pure silicon is converted to *P-type* by introducing an additive to the crystal lattice which has *fewer* valence electrons than silicon. Indium and galium have three valence electrons. These are known as accepter atoms. When they are added to the pure silicon crystal structure, an electron deficiency results. There are holes where electrons are supposed to be. The covalent bond is disturbed. Each indium or galium atom has one less atom than silicon; covalent bonds cannot be formed with silicon

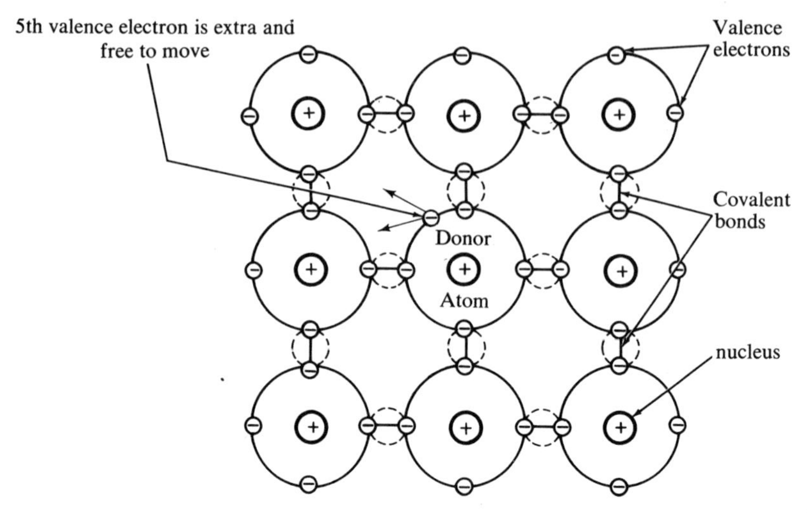

Figure 11–3

atoms. There is no longer a one-to-one situation. Since a hole is an absence of an electron, each hole will accept an electron from a neighboring atom, leaving a hole at the neighbor (Figure 11–4).

To summarize, when pure silicon is chemically treated with an additive which donates electrons, the silicon becomes N type. When silicon is treated with an additive that accepts electrons, the silicon becomes P type.

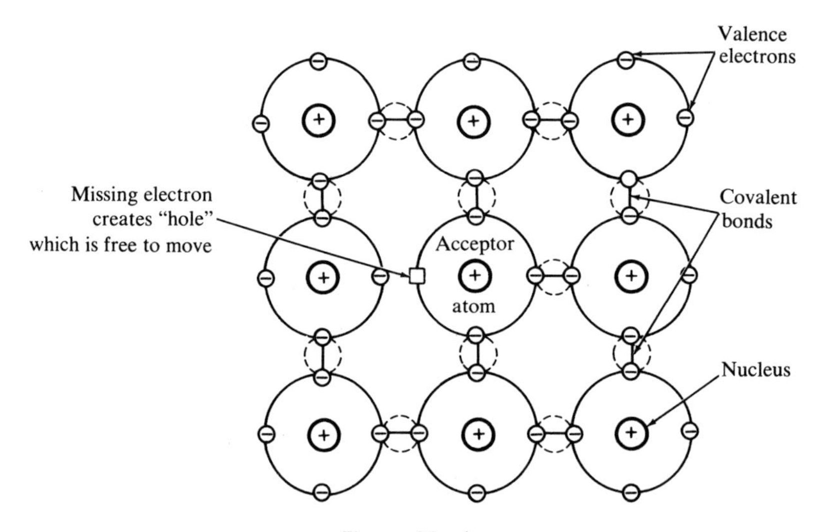

Figure 11–4

Basic Transistor Theory

The surface where P-type and N-type crystals are joined is called a junction. How they join is of little importance to our study of transistors. What happens at the junctions, however, is fundamental (Figure 11–5).

At the junction of the two materials, free electrons in the N material move across the joining surface to fill holes in the P material, as shown in Figure 11–5(b). Some atoms in the N material lose electrons. These atoms become positive ions. Likewise, as shown in Figure 11–5(c), some atoms in the P-type material gain an electron. These atoms become negative ions. Ions build up at the junction, stopping the action because the positive ions in the N material repel the positive holes in the P material. The negative ions in the P material repel the negative electrons in the N material. This process continues until it is impossible for further electron-hole recombinations to occur, as shown in Figure 11–5(d). This junction now has a *space charge*. This space charge voltage is about 0.5 volt. There is no further junction activity until an external voltage, having a greater voltage than the space charge, is applied across the device.

The positive ions form one pole of a battery; the negative ions form the other. When an external voltage is applied with a polarity that has the positive pole connected to the P material and the negative pole to the N material, current flows through the device. The junction is said to be forward biased. Reversing the voltage polarity stops current. The device is said to be reverse biased [(Figure 11–5(e)].

When forward biased, the negative pole of the source repels electrons toward the junction. The positive pole repels holes toward the junction. Once the space charge is overcome by the rush toward it, there is mass recombination of holes and electrons which, effectively, eliminates the junction. Under this condition, the device is a good conductor offering little resistance to current flow from the source, through the device, and back to the source. When reverse biased, the negative poles of the source attract the positive holes in the P material away from the junction. The positive pole of the source attracts electrons in the N material away from the junction. With the junction vacated in this fashion, its effective space charge increases. The junction is a high-resistance, open circuit through which no current can flow, as shown in Figure 11–5(f).

In summary, forward bias pushes holes or electrons toward the junction. Reverse bias pulls holes and electrons away from the junction.

When three semiconductor materials are joined, a transistor is formed and the action changes. The junction behavior remains the same but the

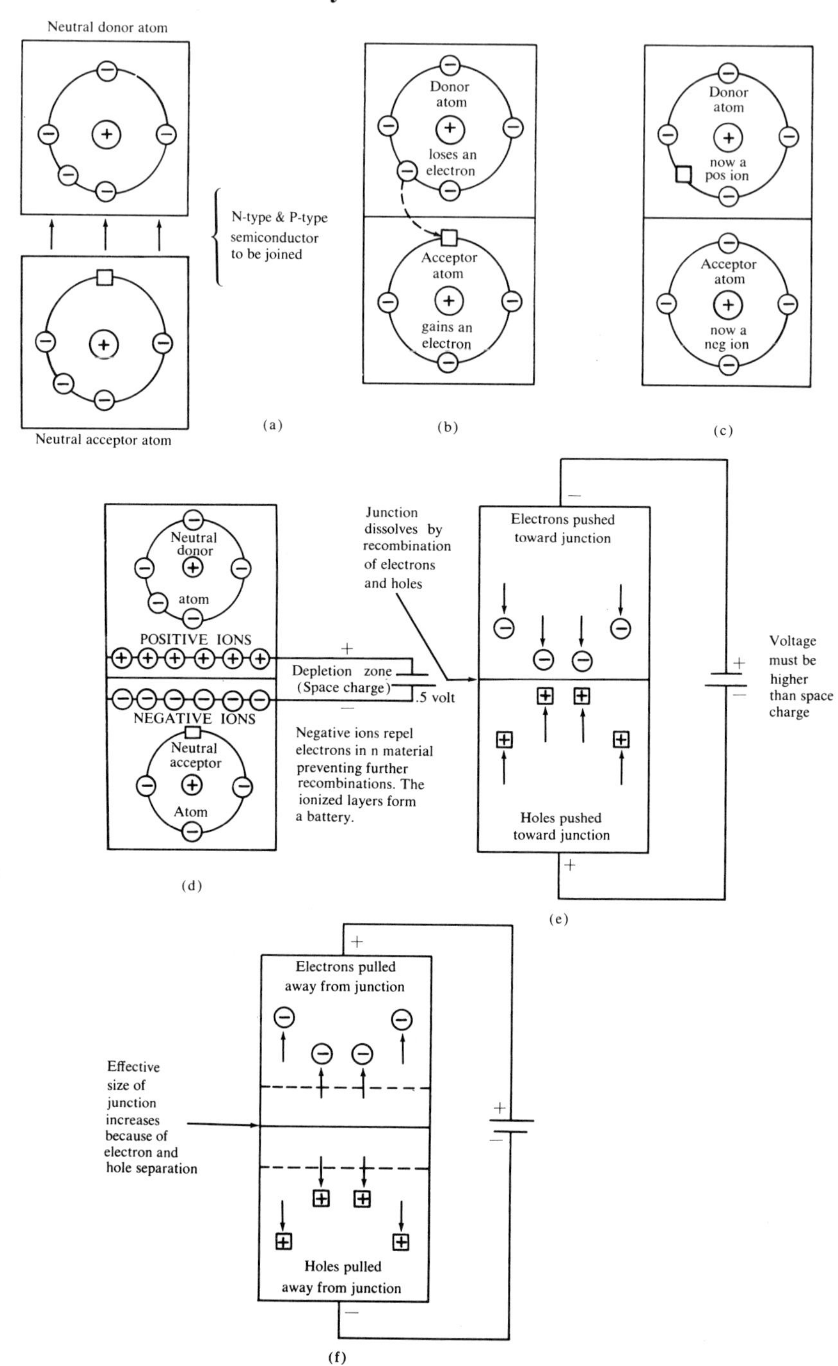
Neutral donor atom
N-type & P-type
semiconductor
to be joined
Neutral acceptor atom
(a)
Donor
atom
loses an
electron
Acceptor
atom
gains an
electron
(b)
Donor
atom
now a
pos ion
Acceptor
atom
now a
neg ion
(c)
Neutral
donor
atom
POSITIVE IONS
NEGATIVE IONS
Neutral
acceptor
Atom
Depletion zone
(Space charge)
.5 volt
Junction
dissolves by
recombination
of electrons
and holes
Negative ions repel
electrons in n material
preventing further
recombinations. The
ionized layers form
a battery.
(d)
Electrons pushed
toward junction
Holes pushed
toward junction
Voltage
must be
higher
than space
charge
(e)
Electrons pulled
away from junction
Effective
size of
junction
increases
because of
electron and
hole separation
Holes pulled
away from junction
(f)

Figure 11–5

entire device, as a unit, behaves differently because the device has two junctions.

There are two transistor types, depending upon the sequence in which the materials are joined. The types are *NPN* and *PNP*. The letters represent the function of each section. The first letter refers to the emitter, the second letter to the base, the third letter to the collector.

Transistor symbols are shown (Figure 11–6) with each of the components labeled. We tell if a transistor is NPN or PNP by looking at the emitter arrow, which always points toward the N material. If the arrow points toward the base, N material makes up the base. The transistor is a PNP type. If the arrow points away from the base and toward the emitter, N material makes up the emitter. The transistor is an NPN type (Figure 11–6).

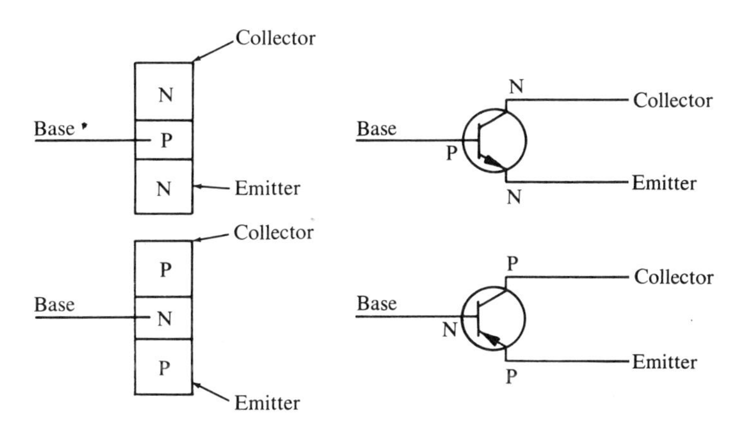

Figure 11–6

In general terms, the emitter/base junction is always forward biased and is the junction across which the input signal is applied. There are exceptions, of course. The collector/base junction is reverse biased. The output signal is usually taken off the collector, but there are exceptions to this. Actually there are three basic circuit forms.

The circuit used most often is the common emitter. The base/emitter junction is forward biased; the collector/base junction is reverse biased. As stated before, the base/emitter forward bias "dissolves" the base/emitter junction. Electrons in the emitter material are pushed toward the base. The base is very thin compared to the emitter and collector. When electrons from the emitter enter the base area, the influence of the collector voltage is felt, and the electrons rush toward the collector which has a high attracting voltage. In other words, electrons are pushed out of the emitter right through this base and are pulled by the collector. Thus, emitter-to-collector current is established. A small emitter-to-base current is also created because some electrons exit at the base.

The base current is quite small in relation to the emitter-to-collector current. Obviously, more current flows through the emitter than the collector because of the additional base current ($I_E = I_B + I_C$). So the ratio of collector current to emitter current (alpha $= I_C/I_E$) is always less than one. Since a small emitter-to-base current controls a larger emitter-to-collector current, there is a gain in current. This is how a transistor amplifies. The larger the emitter-to-collector current compared to the emitter-to-base current, the higher the gain (Beta $= I_C/I_B$). In basic concept, one can look at the forward-biased emitter/base junction as a variable resistor which varies emitter-to-collector current (Figure 11–7).

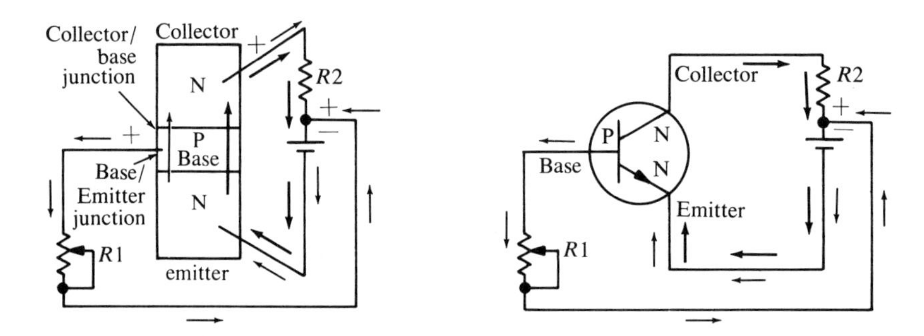

Figure 11–7

The PNP common-emitter circuit operates the same way as the NPN, except that holes are pushed out of the emitter toward the base, dissolving the emitter/base junction. When entering the thin base region, the holes come under the influence of the collector which has a strong attraction for holes. Some holes exit at the base (Figure 11–8). Electron-flow concepts can be used in the PNP device, too. Because holes are positively charged and move toward a negative voltage and electrons are negatively charged and move toward a positive voltage, it is easy to see that holes and electrons move in opposite directions. Electron flow can be traced through a PNP transistor from collector-to-emitter (which is opposite the direction in which holes travel). Whether hole-flow or electron-flow theory is applied is not important. The results are the same even to the polarity of voltage drops across resistors in the collector, emitter, and base circuits. Remember that electrons flow in the external leads that connect the transistor to the other circuit components and battery, whether it is NPN or PNP.

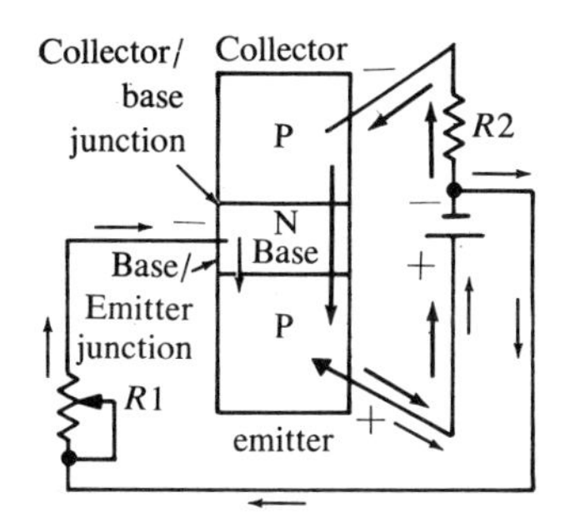

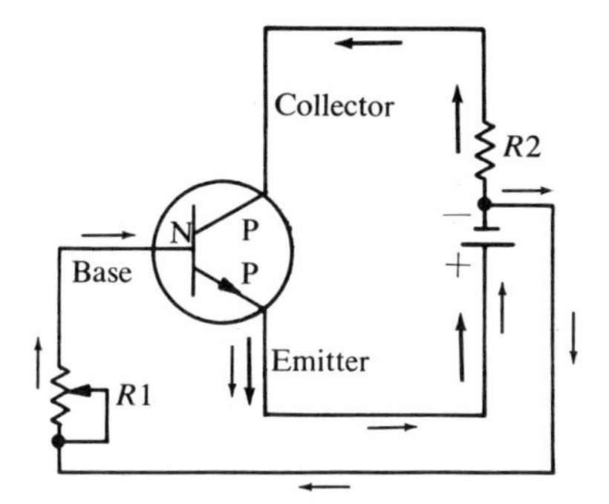

Figure 11–8

In addition to the common emitter, there are the common-base and common-collector configurations. The common base shows the emitter/base junction to be forward biased. Nothing happens in any transistor until the emitter/base junction is dissolved. No matter where a signal is applied or removed, sometime, somehow, the emitter/base junction must be forward biased to cause emitter-to-collector current, as shown in Figure 11–9(a).

In the common base setup, the emitter/base junction is forward-biased. The signal is applied to the emitter and is seen amplified in the same phase (wave is identical) at the collector, as shown in Figure 11–9(b).

In the common-collector circuit, the emitter/base junction is forward-biased. The signal is applied to the base and is seen at the emitter in the same phase, as shown in Figure 11–9(c). (The common emitter is the only circuit that features signal-phase inversion.)

Each of these three basic circuit patterns has special features, but the transistors all operate in the same way, as shown in Figure 11–9(d). The only difference is where signals are inserted and removed.

Summary

Common emitter input is from base to emitter. Its output is from collector to emitter. This configuration has voltage and power gain.

Common base input is from emitter to base and its output is from collector to base. This configuration provides voltage gain and has medium power gain.

Common collector input is from base to collector and its output is from emitter to collector. Its characteristics are current gain and no voltage gain. It has a small power gain. This configuration is used mainly as a method of matching impedances. Impedance matching provides circuit efficiency.

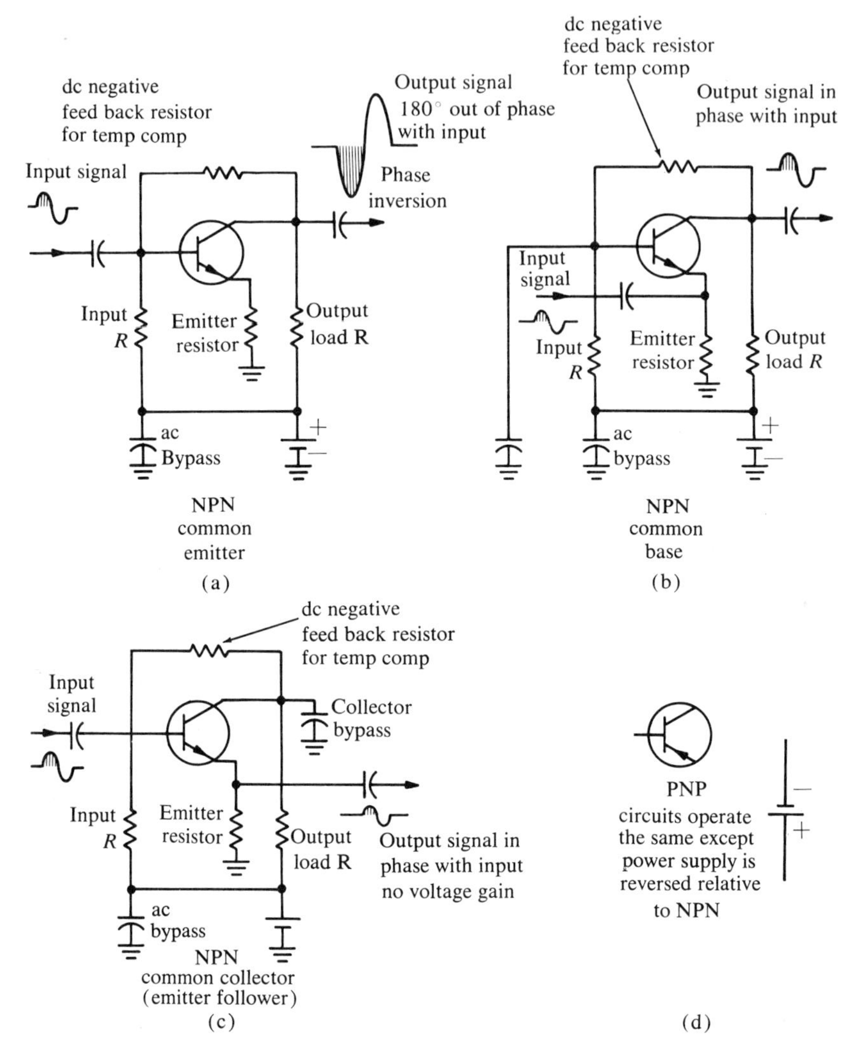

Figure 11–9

Practical Side of Transistors

A technician need not be a physicist to understand how a transistor functions (nor to recognize one that does not). A few basic ideas, learned well, will do. A brief review is included here as an aid to understanding transistors from a practical viewpoint.

An NPN transistor behaves very much like a triode vacuum tube as far as voltage polarities are concerned. Forward bias is required to make an NPN transistor turn on (conduct). Reverse bias turns off (cutoff) an NPN transistor. Forward bias makes the base positive, as seen from the emitter, by about 0.7 volt positive dc in a silicon transistor. Much less than this will allow the transistor to cut off. For some circuit uses, a negative reverse bias may be necessary to achieve the circuit goal.

The letters N, P, and N tell us much. The center letter, P, tells us the base must be positive with respect to the emitter in order to turn on the (forward bias) transistor. The first two letters, NP, tell us that the negative lead of a voltmeter should be on the emitter when one is measuring forward bias, with the positive lead on the base. The letters PN (base/collector) must be reverse-biased. This means that the collector (N) is tied to a positive voltage and the base (B) is tied to a negative voltage. The more the base voltage is increased in the direction of the collector polarity, the heavier the transistor conducts. The more the base voltage moves in the direction of the emitter, the less the transistor conducts (until it actually stops completely).

A conducting transistor is one in which a low resistance is connected across the power supply. A saturated transistor is one in which an increase in forward bias cannot increase collector current. A saturated transistor is one in which there is very low resistance—almost a dead short. A cutoff transistor is considered to have a high resistance as far as dc is concerned and can be termed "out of the circuit."

Figure 11–10 shows correct voltages for an NPN circuit. The collector voltage is lower than its effective source, indicating current through *R*2. The base voltage is lower than the effective base source, indicating

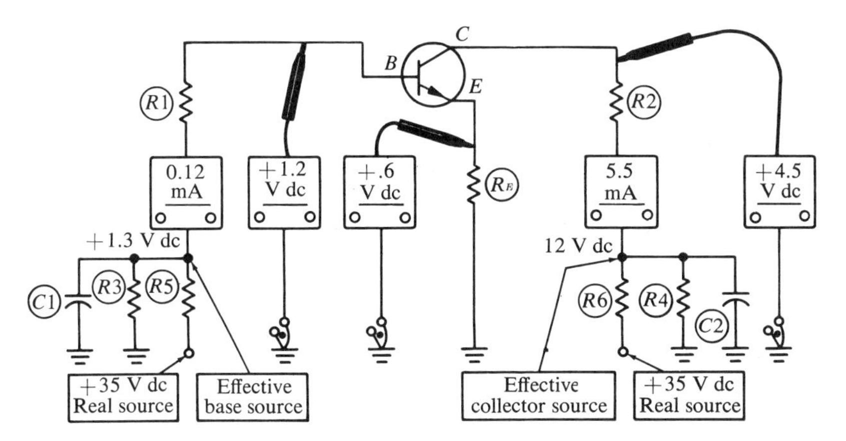

Figure 11–10

current through $R1$. Also, voltage at the emitter indicates current through R_E. The stage is not saturated as indicated by the collector voltage which is slightly less than half the source value (12 V dc). Note the base/emitter forward bias of 0.6 V dc, with the base positive with respect to the emitter. This represents a forward-bias condition. The collector is the most positive element in an NPN device.

One final aid is the transistor symbol. The arrow always points away from the positive power-supply voltage and in the direction of the negative supply terminal. This is true whether the transistor is an NPN or a PNP type.

Correct voltages for a PNP-type circuit are shown in Figure 11–11. Forward bias is required to make a PNP transistor turn on (conduct). Reverse bias turns a PNP transistor off (cutoff). Forward bias makes the base negative, as seen from the emitter, by about −0.7 V dc in a silicon transistor. Much less than this will allow the transistor to cut off.

The letters *PNP* again tell us what we need to know about polarities. The center letter, N, tells us the base must be negative with respect to the emitter in order to turn on (forward bias) the transistor. The letters PN tell us that the positive lead of a voltmeter must be on the emitter when one is measuring forward bias, with the negative lead on the base. The letters NP (base/collector) must be reverse biased. This means that the collector (P) is tied to a negative voltage and the base (N) is tied to a positive voltage. The more the base voltage is increased in the direction of the collector polarity, the heavier the transistor conducts. The more the base voltage moves in the direction of the emitter, the less the transistor conducts, until it is cut off completely.

A conducting transistor has low resistance across the power supply. A saturated transistor is practically a dead short. A cutoff transistor has high resistance and can be calculated to be out of the circuit. The PNP arrow symbol for the emitter points away from the positive power-supply terminal in the direction of the negative power-supply terminal. The arrow points to the N material in every transistor.

Figure 11–11 shows collector voltage lower than its effective source, indicating current through $R2$. Likewise, base voltage is lower than its effective source, indicating current through $R1$. Also, voltage at emitter indicates current through R_E. The stage is not saturated, as evidenced by a collector voltage slightly less than half the source (−12 V dc). Note the base/emitter bias of −0.7 V dc, with the base more negative than the emitter. This represents a forward-bias condition. Circuit action is similar to that of NPN, except all voltages are reversed in polarity. Also electron-current flow is from collector to emitter.

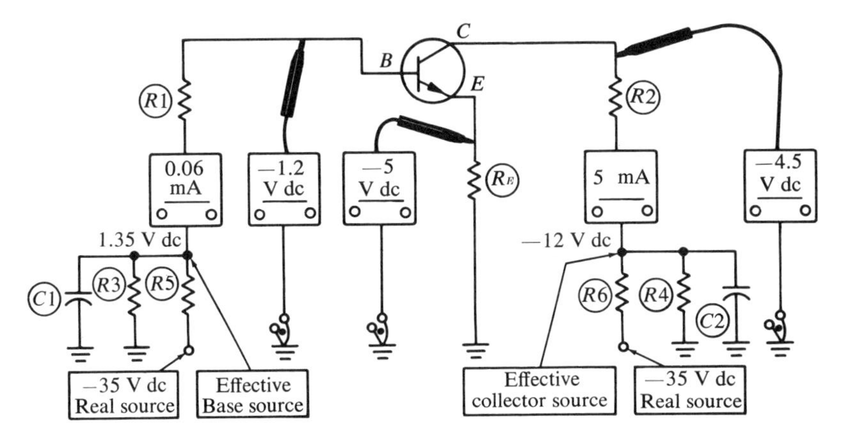

Figure 11–11

Figure 11–12 shows the same circuit shown in Figure 11–11, but with a negative ground power supply. This technique enables both NPN and PNP devices to operate from a common positive output power supply. PNP transistors require negative collector voltage (as seen from the emitter). This is achieved by connecting the positive source voltage to the emitter, thus making the collector negative, relative to the emitter. By the use of resistors we can divide the voltage so that the base is made 17 V dc negative, as seen from the emitter. Notice the voltage drop across $R1$, $R2$, and R_E, indicating current through these components.

That the stage is not in saturation is shown by the difference between the collector and emitter. The IR (voltage) drops across $R2$. The tran-

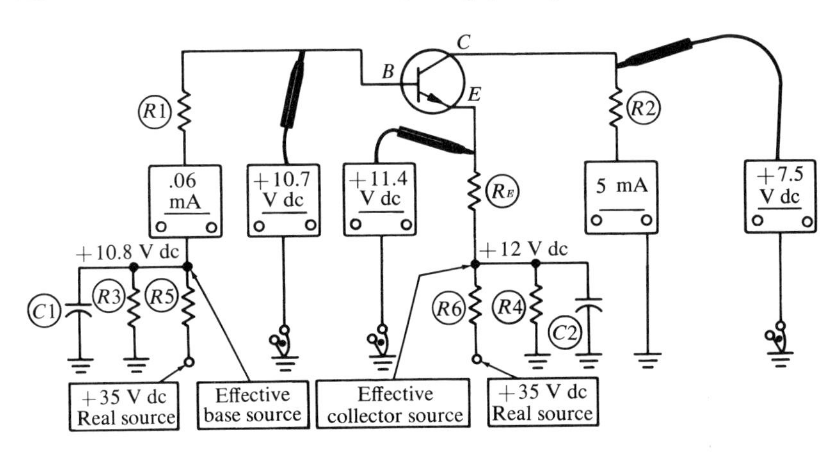

Figure 11–12

sistor and R_E add up to the 12 V dc source. This is a basic Ohm's law demonstration for series circuits.

Now let us determine the meaning of incorrect voltages. Forward bias enables the emitter of the transistor to release electrons just as a heater in a vacuum tube enables the cathode to release electrons. This may seem oversimplified but is sufficient for our purposes. Forward bias for a transistor replaces the heater in a vacuum tube. Just as an absence of heater voltage renders the vacuum tube inoperable, an absence of forward bias at the transistor's base/emitter junction renders the device inoperable.

As a first step, service technicians have always looked for a lit tube. It is an equally good technique to check a transistor for forward bias as a first step.

Base voltage is established by divider similar to $R1$, $R5$, and $R3$ in Figure 11–13 which shows an NPN transistor. An open $R1$, $R5$, or a shorted $C1$ will remove the base from its source. With low or missing base voltage, the transistor cuts off. The collector rises to the supply value. With no collector current, no voltage drop can appear across the emitter resistor, R_E.

This is an easy problem to locate. No voltage on the base and emitter and a high collector voltage tells us the stage is cut off by the absence of forward bias. A good rule to learn here is that abnormal base voltage, when measured with respect to ground, is caused by a defect in the base circuit itself.

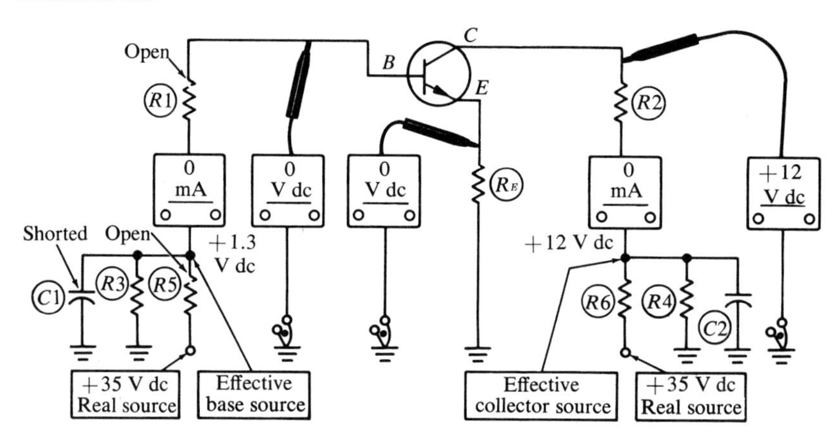

Figure 11–13

An open $R3$ in the base divider network will permit base voltage to increase, turning on the transistor heavily. This results in a low collector voltage because of the IR drop across $R2$. The transistor might destroy itself by heavy conduction. Notice that the rule described above still applies.

Similar defects in a PNP circuit (Figure 11–14) produce dissimilar voltages, but the basic rule described still applies. An open at $R1$ or $R5$ will allow the base to go toward the emitter because of the low base-to-emitter internal resistance. A positive 12 V dc on the emitter is coupled to the base through the low internal resistance. The fact that the emitter is at effective source voltage (12 V dc) indicates there is no current through R_E (the emitter resistor) to cause a voltage drop. With no current through R_E, we assume the stage to be cut off. We further assume this to be caused by a lack of forward bias which can be verified by measuring the base-to-emitter voltage. Finally, because the base voltage is far from normal, as measured from ground, trouble is indicated in the base circuit itself. An open $R3$ will not permit the base to be negative with respect to the emitter. The stage will cut off, producing a reverse-bias situation.

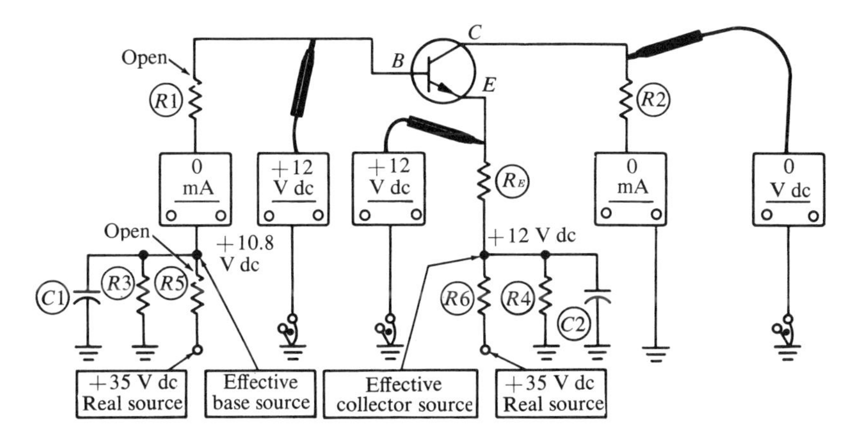

Figure 11–14

As in Figure 11–15 let us assume R_E, the emitter resistor, is open. The base voltage, measured from ground, is normal. We can conclude the defect is not in the base circuit. The measuring of the emitter voltage with respect to ground could lead us to the erroneous conclusion that the stage is conducting. We can tell if the stage is conducting by taking a reading of the collector voltage which, in this instance, is at the effective source value (12 V dc). (We know that this cannot be the case when the transistor conducts.) The voltage on the emitter is coupled from the base via the low base/emitter resistance. Checking forward bias measured from the emitter reveals the final clue. The meter will read zero volts dc. The base voltage seen with respect to ground is normal. We assume the base circuit to be normal. The stage is cut off by an open emitter resistor. Remember with the stage cut off, the collector sits at effective source value. There is no voltage drop across $R2$. Also, an open R_E prevents base current. Accordingly, there is no voltage drop across $R1$.

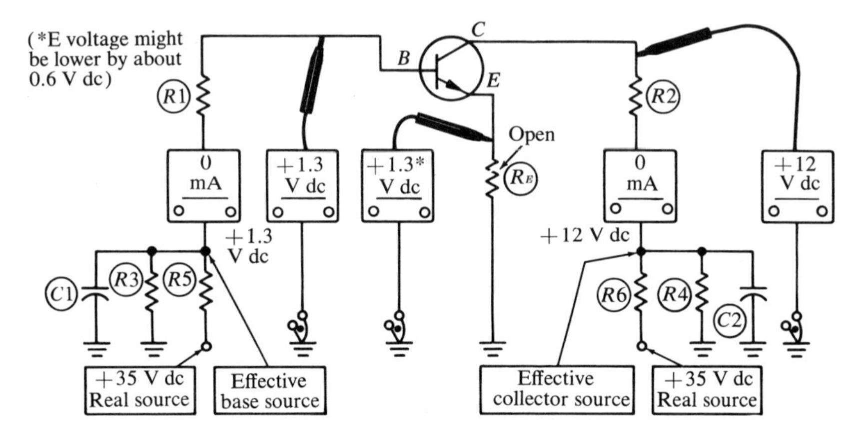

Figure 11–15

An open emitter resistor (R_E, Figure 11–16) removes forward bias, preventing device conduction. The emitter assumes the base voltage via internal resistance of the transistor. Some transistors may exhibit a potential difference between base and emitter which could cause misleading forward bias readings. The actual circuit condition is indicated through a reading of the voltage between the elements. In this instance there is no voltage drop across $R2$, indicating there is no current passing through it. That there is no current passing through the base circuit, is indicated by lack of voltage drop across $R1$.

An open $R4$ would raise the effective source from 12 V dc. Since collector current is independent of collector voltage, there is no significant change in collector current. Collector current is a function of base current. The defect may not reveal itself if the device can stand high emitter/collector voltage.

An open collector in the NPN circuit of Figure 11–17 presents some confusing effects. With an open collector load resistor, $R2$, there is no path for completing the collector circuit. No voltage drop (current × resistance) appears across $R2$. There is base current producing an appropriate voltage drop across $R1$. Base voltage is normal. The small emitter voltage is caused by small base current through R_E.

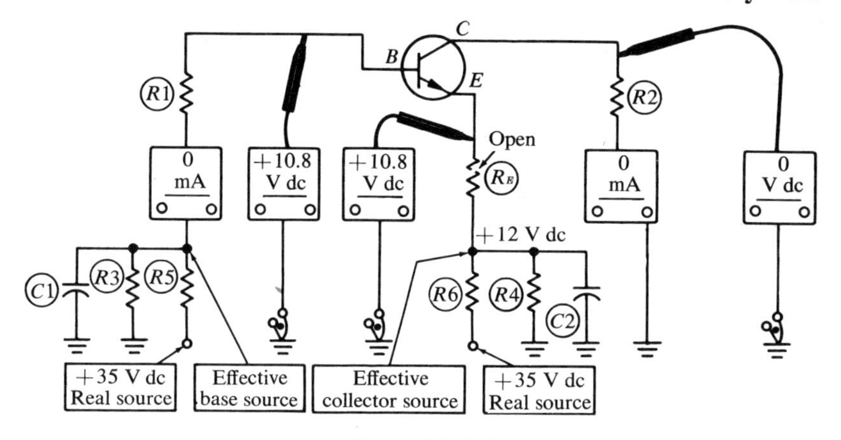

Figure 11–16

The normally reverse-biased collector/base junction becomes forward biased in this situation, resulting in the assumption of the emitter voltage by the collector because of the low resistance collector/base junction. The 0.06 V dc on the collector may be puzzling. When the collector voltage drops below the base, the collector/base junction becomes forward-biased. The collector assumes the emitter voltage in this case.

An open collector circuit exists when all voltages read close to each other, with the base voltage normal. The base current is higher because the base is the sole element attracting electrons from the emitter. A shorted $C2$ produces the same readings because it grounds the lower end of $R2$, removing collector voltage.

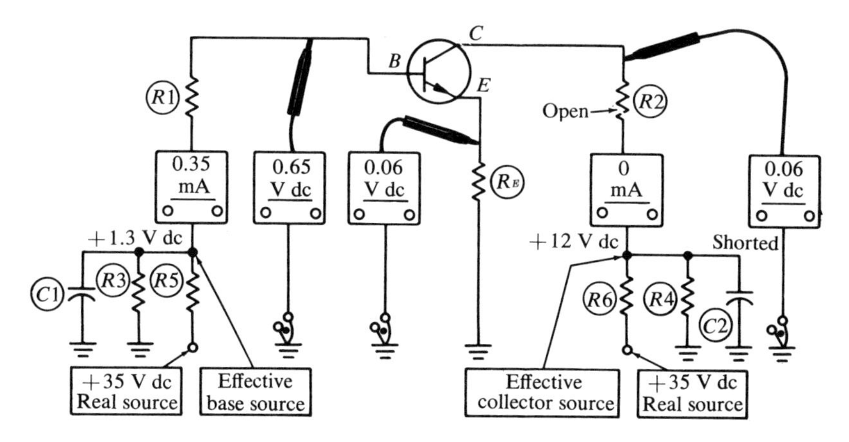

Figure 11–17

Figure 11–18 demonstrates collector-circuit defects showing incorrect voltages and currents of a PNP transistor. An open collector load resistor, R2, prevents collector current because there is an incomplete circuit. With no current, there is no voltage drop (current × resistance) across $R2$ (although the 11.9 V dc read across the open resistor can be misleading). Not knowing the resistor is open, and reading the 11.9 V dc across it, would lead to the erroneous conclusion that current must be passing through $R2$. This shows the importance of reading all the voltages on a transistor. In this instance, the voltage at the collector is the emitter voltage taken by the collector as a result of the forward-biased collector/base junction. The base voltage is close to normal. The small voltage drop across R_E is due to base current through it. Forward bias from base-to-emitter is a normal 0.6 V dc with the base negative, as seen from the emitter. Base current has risen from a normal 0.06 mA to 0.35 mA because the base is the only element attracting electrons from the emitter. When all voltages read practically the same, and the base voltage is normal in the PNP circuit, we have an open collector circuit.

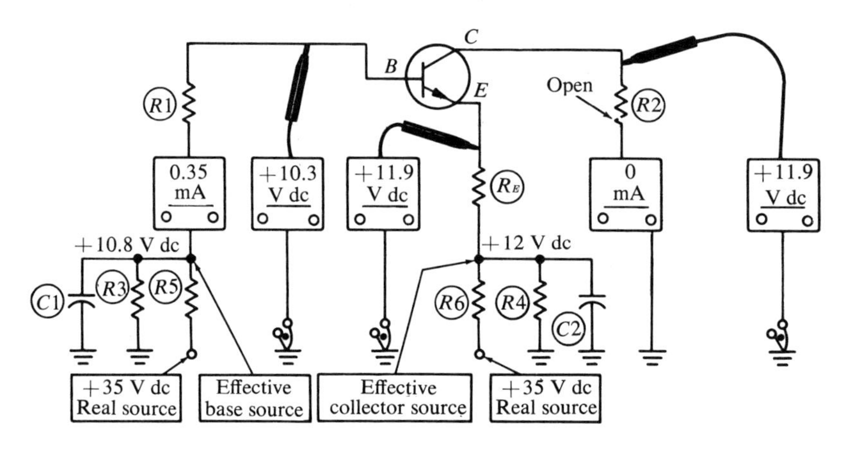

Figure 11–18

At this point let us study the effect of transistor defects on circuit measurements. It is easy to be misled when an open is inside the transistor as compared to an open outside the unit. With an open base lead inside the unit, the stage will cut off. No current will flow to develop any voltage across the emitter resistor, R_E (Figure 11–19). The collector rises to source value 12 V dc, further indicating a cutoff stage. Yet forward bias as seen from the emitter is unusually good. Measured from ground, the base voltage is normal, ruling out the external base circuit which is still connected to its voltage divider. The fact that the emitter has no voltage reveals there can be only infinite base/emitter resistance (an open base or emitter lead inside the unit). An open circuit is considered to have infinite resistance.

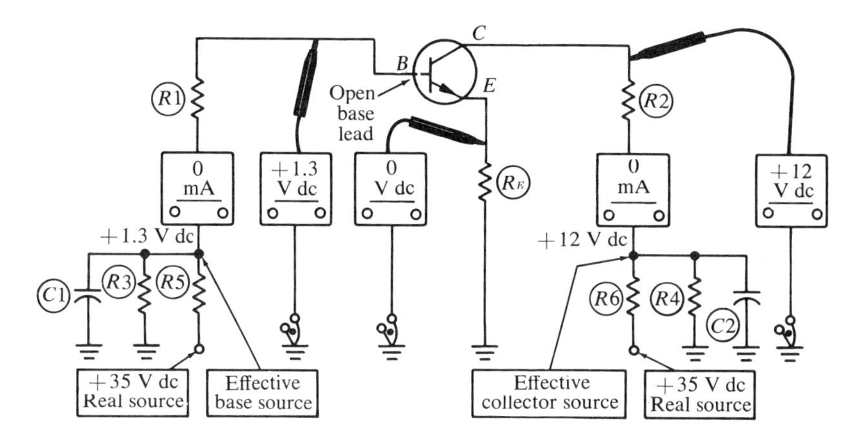

Figure 11–19

The PNP transistor in Figure 11–20 produces similar effects when base/emitter resistance is infinite. Unusually good forward bias read from the emitter may fool us into thinking the stage must conduct. Base voltage with respect to ground is nearly normal, which rules out external base circuit defects. We know the stage is open despite the exceptional forward bias because there is no voltage dropped by R_E. Also, there is no voltage dropped across $R2$. We conclude that when a stage has unusually good forward bias, but no collector current through the load or the emitter resistor (or any indication of conduction), the transistor itself must be suspected of having an open base or emitter lead inside the unit (refer to Figures 11–20 and 11–22).

Figure 11–21 shows a transistor with an open emitter lead of an NPN type. Because there are no voltage drops, there is no base current and also no collector current. If the defect were an open R_E, the emitter would assume the base voltage. With 1.3 V dc on the base, and 12 V dc on the collector, it is safe to assume that $R1$ and $R2$ are operable. The voltages are available and the components check out according to the rules. There is nothing left in the circuit but the transistor, which has an open emitter lead internally.

Figure 11–22 shows a PNP transistor with an open internal emitter lead. Full source voltage is on the emitter lead. There is zero voltage drop across $R2$ and $R1$, confirming lack of current anywhere. Voltage on the emitter and base rules out open resistors in those leads. An open

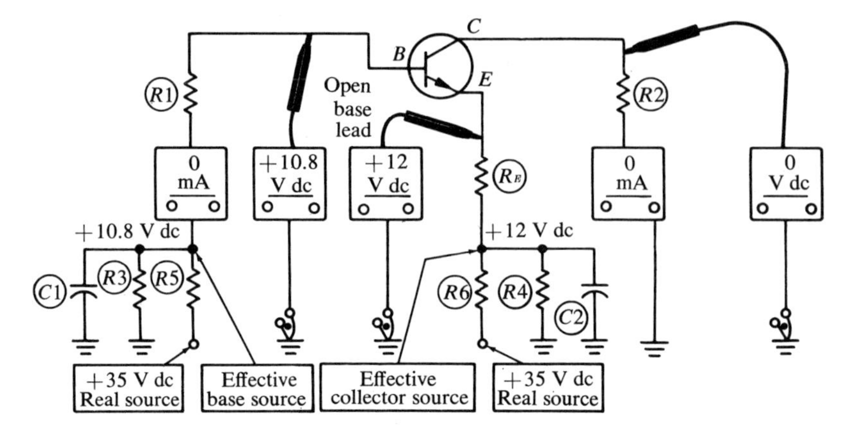

Figure 11–20

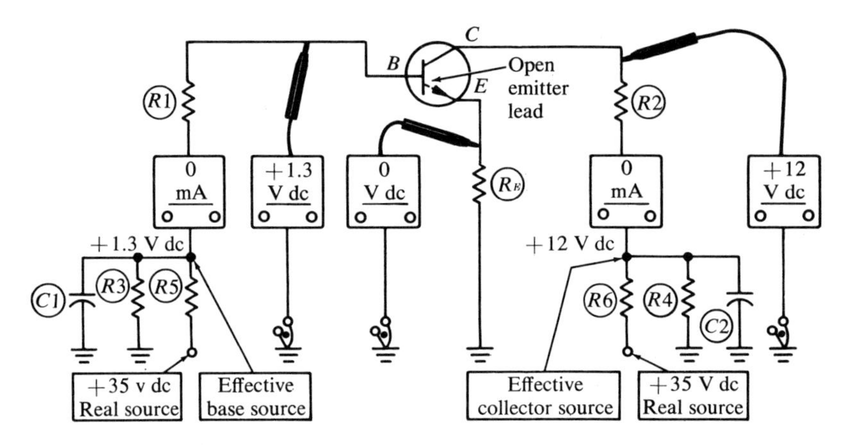

Figure 11–21

$R2$ would allow some voltage on the collector picked up from the emitter inside the unit. These checks establish that the components are in order and voltages are available. The device is defective due to an open emitter lead internally.

Now we come to the last element of the transistor, the collector. An open collector lead inside the transistor will produce normal base voltage read from ground. Forward bias will be normal but not exceptionally good or bad. Yet the collector voltage on an NPN will rise to the effective source value 12 V dc, indicating the stage to be cut off. Further indication is the lack of voltage across $R3$ and $R5$. A small voltage drop appears across the emitter resistor (R_E) due to the base current passing through it (see Figure 11–23).

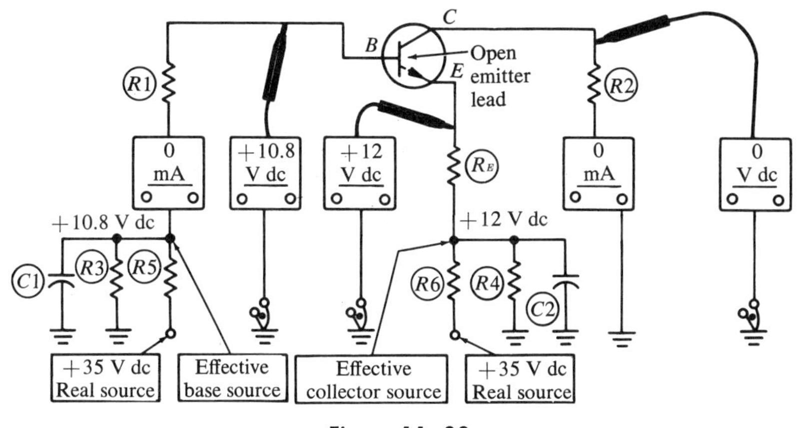

Figure 11–22

Figure 11–23 shows an NPN transistor with an open internal collector lead. The full source voltage on the collector reveals the cut off stage and closed $R2$. Forward bias voltage is normal with base current a little higher than normal, indicating a satisfactory base/emitter circuit. With all voltages present and the stage inoperabe, it must be concluded the device itself is defective. In this instance, the collector lead is open internally.

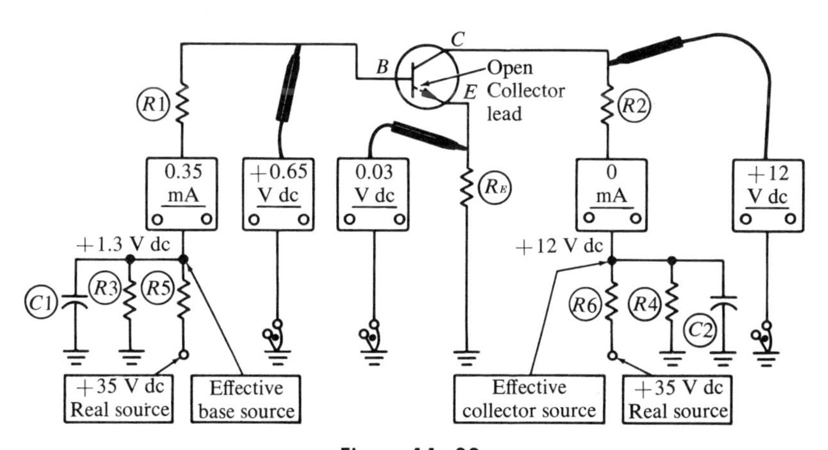

Figure 11–23

Figure 11–24 shows a PNP transistor with an open internal collector lead. There is no voltage drop across $R2$, indicating a lack of collector current through it. An open $R2$ would cause the collector to take on the emitter voltage. Since this is not the case here, a quick check of forward bias assures us that the base/emitter circuit is in good order. A small drop across R_E is due to base current. The logical conclusion here is that the device is open internally at the collector since everything else is operable. When forward bias is normal or unusually good, but the stage does not conduct, suspect an open internal lead.

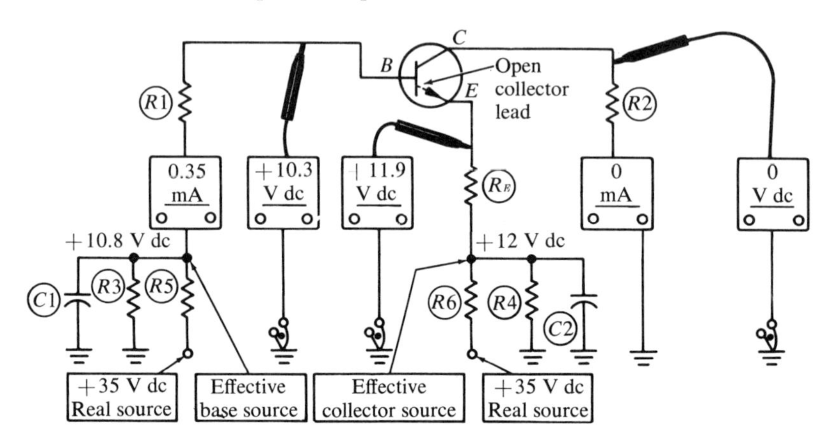

Figure 11–24

Now let us look at transistors that are shorted or have a high leakage characteristic. High current drain or "leaky" transistors are quite common, especially with low-power types. Their voltage readings are dependent upon the extent to which the transistor is defective. In the NPN circuit of Figure 11–25 an emitter-to-collector leakage permits increased collector current, resulting in a substantial *IR* drop across the collector lead ($R2$). The high emitter current through R_E develops a high *IR*

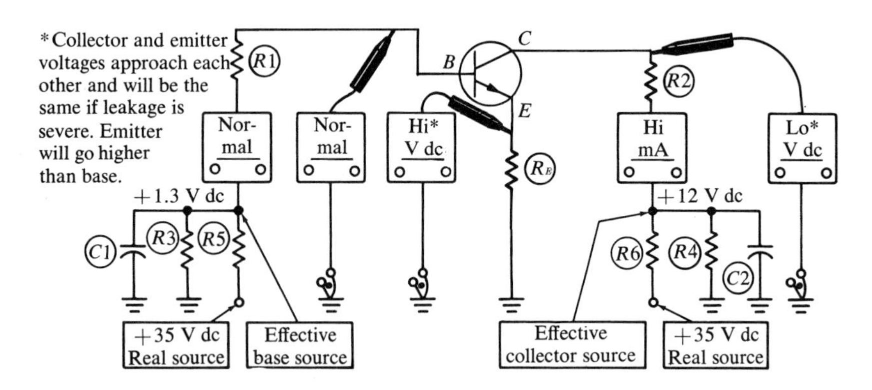

Figure 11–25

drop at the emitter terminal. This IR drop could actually reverse bias the base/emitter junction, yet the stage will conduct heavily. Base voltage is normal (from ground). Shorting the base to the emitter should cut off a transistor, but in this case, collector current continues to be heavy. This indicates a leaky transistor. Collector current cannot be controlled by the base.

The leaky PNP transistor (Figure 11–26) results in heavy collector current with consequent high IR drop across the collector load ($R2$). This causes the collector voltage to read low. Heavy current through R_E drops the voltage at the emitter terminal, perhaps enough to reverse bias the base/emitter junction. Shorting the base to the emitter does not stop collector current through a leaky transistor. Basically, if forward bias is low or zero, but the transistor conducts heavily, we should suspect a leaky transistor.

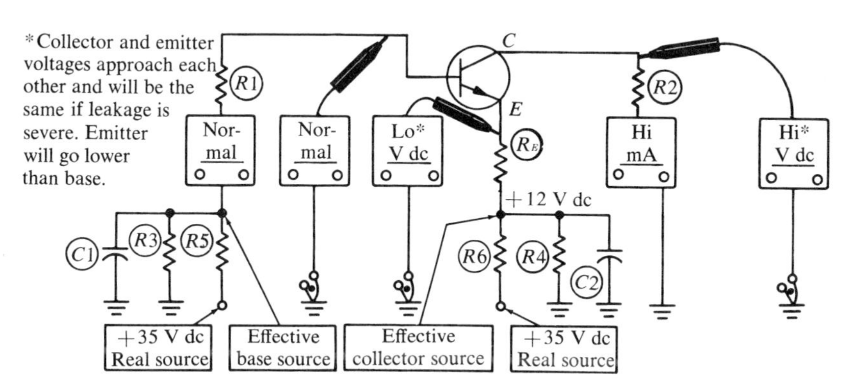

Figure 11–26

When dc voltage checks lead to suspicion of a defective transistor, an ohmmeter reading is necessary. Since a transistor actually consists of two diodes with the base element in common, the forward resistance and reverse resistance of each diode can be measured.

First the base/emitter diode is checked with the ohmmeter's common lead placed on the base, and the other lead on the emitter. Note the resistance reading of the ohmmeter. If the diode is not defective, reversal of the ohmmeter leads will indicate a completely different resistance. In one direction, the resistance is very low. In the other direction, the resistance is high. With most low-power transistors it is best to use the $R \times 100$ scale to prevent excessive current from passing through the diode when the ohmmeter leads bias the junction forward. Higher resistance-scale ranges, such as $R \times 1000$ or $R \times 10{,}000$, cause too much voltage to appear at the leads in some ohmmeter models. The $R \times 100$ scale is an excellent compromise. Using this range, you will find that diode resistance will read low in one direction and very high in the opposite direction.

The base/collector junction must also be checked in both directions. First measure and note the resistance with the common lead of the meter on the base. Then reverse the leads. The resistance should be considerably different. Whether the reading is low or high in any particular direction depends upon the voltage polarity at the ohmmeter leads. The change, or difference, in ohmmeter readings when the leads are reversed is the important consideration. A transistor that reads low in both directions is probably shorted or leaky. One that reads high in both directions is open. A transistor that does not present large differences in resistances when the ohmmeter leads are reversed is probably leaky.

Out-of-circuit ohmmeter checks are more valid than those checks made with the transistor in the circuit, although in-circuit tests are also valid if the serviceman takes all dc paths into account when interpreting the readings. The important point is that a great difference should be noted when the ohmmeter leads are reversed when connected across any one diode—whether the diode is in- or out-of-circuit. Direct current paths in parallel with the ohmmeter result in lower resistance readings, but the difference between forward and reverse directions will be apparent.

Use the $R \times 10$ scale with medium-power to high-power transistors. Again, the forward and reverse resistances must be significantly different (low in the forward direction, very high in the reverse).

Leakage between emitter and collector is simple to detect. Place the ohmmeter leads across the two elements and note the reading. Reverse the leads and read the resistance again. If the resistance is low in either or both directions, the transistor is leaky or shorted. The resistance should be high in both directions (about 5,000 ohms or more with low-power types). On the $R \times 100$ scale, the unit might appear to be open. Medium-power to high-power types will read lower to about 500 ohms or more. On the $R \times 100$ scale, these should appear to be open. The base lead is disregarded in making the emitter-collector leakage test. These tests are not practical when made while the transistors are in the circuits.

questions

1. Which of the following phrases is correct?
   a) Beta-change in collector current + change in base current.
   b) Beta-change in base current + change in collector current.
   c) Beta-change in either current + change in collector current.
   d) Beta-change in base current + change in either current.
   e) None of the above.
2. Check off two incorrect phrases.
   a) The P-type collector of an NPN transistor, etc.
   b) The N-type base of a PNP transistor, etc.
   c) The P-type emitter of a PNP transistor, etc.
   d) The N-type emitter of a PNP transistor, etc.
   e) The N-type collector of a PNP transistor, etc.
3. Check off the equivalent pair.
   a) Plate-Base
   b) Plate-Emitter
   c) Cathode-Emitter
   d) Base-Cathode
   e) Emitter-Grid
4. The collector current of a grounded emitter amplifier is controlled by:
   a) Emitter current
   b) Base current
   c) Collector voltage
   d) Battery voltage
   e) Voltage across RL
5. Signal voltage output of a grounded emitter stage is obtained at a collector by connecting:
   a) A resistor in the base circuit
   b) A resistor in the collector circuit
   c) A resistor in the emitter circuit

d) A resistor between collector and base

e) None of the above.

6. In the grounded collector stage the following electrode is common:

 a) Base

 b) Emitter

 c) Grid

 d) Collector

 e) None of the above

7. Fixed bias is obtained by connecting a resistor between:

 a) Collector and emitter

 b) Collector and ground

 c) Emitter and ground

 d) Collector and base

 e) Voltage supply and base

8. In a PNP transistor a negative-going signal applied to the base of a grounded emitter amplifier will:

 a) Increase base current

 b) Decrease base current

 c) Decrease collector current

 d) Decrease emitter current

 e) None of the above

9. In a grounded emitter amplifier, amplifier-bypassing the emitter with a condenser will prevent:

 a) dc negative feedback only

 b) ac negative feedback only

 c) Both dc and ac negative feedback

 d) Temperature rise

10. The collector voltage with a resistant load can never be greater than the voltage supply. True _____ False _____

11. A grounded collector stage provides:

 a) Phase inversion

 b) High current gain

 c) High voltage gain

 d) Very high power gain

 e) None of the above

12. Draw a common emitter stage (PNP) with fixed bias. Bias voltages must be correct.
13. Draw a common emitter stage (PNP) with self-bias. Bias voltages must be correct.
14. Repeat Problem 12 using an NPN transistor.
15. Repeat Problem 13 using an NPN transistor.

12 the receiver

Imagine the serviceman who walks into a customer's home and finds he can't turn on the receiver. Does he turn the "on-off" switch from a counterclockwise position to a clockwise position? Does he push the knob or pull it? If he doesn't know, he will end up with the knob in his hand and egg on his face. What could be worse than the serviceman who has to ask the customer to turn on the set for him? When a serviceman loses the customer's confidence, he loses the customer.

The importance of knowing the receiver layout prior to a service call cannot be overstressed. Most major television manufacturers provide excellent service manuals written by excellent technical writers. You should use them to gain much-needed mechanical and electrical information on the specific receiver you will be servicing.

Chapter 12 is designed to familiarize the student with the service manual, mechanical terminology, and basic, often-used tools. Again, Motorola's TS-915 is the teaching tool. But it is only that. The following chapter should help prepare you to read and understand any service manual and to understand the mechanics of in-home servicing for any TV receiver.

Panel Removal

It is just as important to know how to remove the receiver chassis or panels as it is to know how to turn a set on and off.

All panels can be removed without the aid of a soldering iron. Most can be removed without tools.

Because of the need for good RF and dc grounding, the following panels incorporate ¼-inch screws in their mounting (a 1-inch spintite is required for their removal): the color panel (S, Figure 12–3), video-IF panel (B, Figure 14–2), and video-output panel (M, Figure 12–3).

In some models the horizontal panel (F, Figure 14–2) is secured for shipping purposes by the insertion of armite wedges that are placed through the plastic mounting studs. These may also be discarded. Any wires that plug directly onto the panel must be removed by pulling straight out. *Do not pull at an angle.*

Most panels can be taken off their nylon mounting studs with the fingers. Apply just enough pressure at the studs to free the panel from them. Then, lift the panel off the connectors. When panels are not readily accessible by hand, they may be pried off their mounting studs. *Extreme care should be taken to avoid fracturing panels.* Use a flat tool (preferably insulated) applied between the panel and chassis. Lift the panel off the studs by applying pressure at its edge and adjacent to the studs, only. Free each stud, then lift the panel off the connectors by hand.

Panel Installation

Position the replacement panel on nylon mounting studs. Make certain panel-connecting plugs are aligned with chassis-connecting plugs. Use a dull flat tool, preferably a hollow tool such as a spintite wrench. Place the spintite wrench over the nylon mounting stud and press the panel into position. On panels that are not readily accessible with a spintite wrench, place a dull flat tool adjacent to the mounting stud (away from the corner of the panel) and press into position. Make certain all panel screws are in place and tight to insure proper grounding of the panel. Replace all connections and make the necessary adjustments. For detailed instructions pertaining to each panel, see the instructions that are packaged with the replacement panel.

Transistor Replacement

When replacing any "plug-in" transistor (i.e., the horizontal outputs, vertical output, horizontal current regulator, and audio output), please observe the following precautions:

1. The transistor sockets are not "captive," that is, the transistor mounting screws also secure the socket. When one installs the transistor, the socket must be held in its proper location. This location is indicated by flanges on the socket which fit into the chassis or heat sink.
2. When one replaces the horizontal output transistors *silicon grease* should be applied evenly to the area between the transistor and the heat sink.
3. When one replaces any other "plug-in" transistor, silicon grease should be applied evenly to the area between the transistor and the mica insulator. The silicon grease acts as a conductor for the transistor-radiated heat.
4. Both transistor mounting screws must be used. In some applications only one screw is used to make electrical contact.
5. All transistor mounting screws must be tight before applying power to the receiver. This insures proper cooling and electrical connections. Note: Use caution when tightening transistor mounting screws. If the screw threads are stripped by excessive pressure, a poor electrical and mechanical connection can result. *Noncompliance with the above instructions can result in failure of the transistor and/or its related components.*

Chassis Removal

See Figures 12–1, 12–2, and 12–3 for the location of screws and latches. *Do not push or pull on plug-in panels during removal.* Remove two chassis retaining and one ac interlock screw at the rear of the cabinet. Remove the cabinet back cover. Remove the video-output panel (M) from its mounting at the top of the cabinet. Disconnect all interconnecting cables from the chassis to the cathode-ray tube (CRT) and the power supply. Slide the chassis forward and depress the retaining latch located at the front of the lower chassis track.

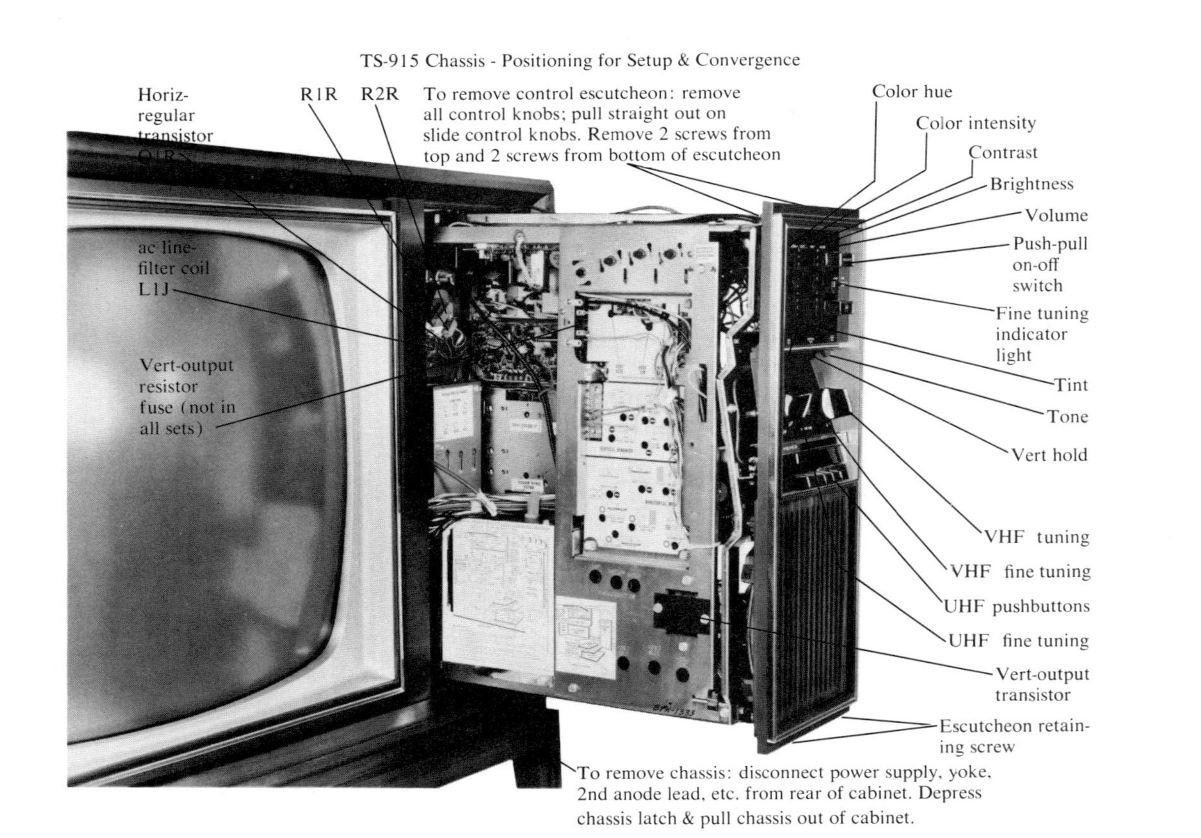

Figure 12–1
Courtesy of Motorola, Inc.

The Circuit Guard

The circuit guard is a thermal cutout relay, connected in series with the power-transformer primary. Excessive heat caused by an overload will open the circuit-guard contacts. This removes ac voltage from the power transformer.

The circuit guard has a trip-free design, allowing power to the set only when the reset button is released. The circuit guard cannot be defeated by holding in the reset button.

Alternating Current Line-Filter Fuse

An ac line filter consisting of a choke (*L*1J, Figure 14–2 and foldout schematic) and a 0.15-mF capacitor (*C*1J, Figure 14–2 and foldout schematic) is connected in the primary circuit of the power transformer. An ac line-filter fuse consisting of a ⅝-inch piece of #31 wire has been inserted in series with *C*1J to protect the ac line source from damage if this capacitor were to short. *The receiver will operate with this fuse open.* It is suggested that this fuse be checked during normal service to insure proper ac line filtering.

Bezel Removal

See Figure 12–2 for the location of bezel retaining screws. An 18 × ¼-inch spintite will expedite bezel removal. Chassis TS-915 has six screws that can be removed from the inside of the cabinet.

Note: For easy access to the lower-left bezel screw, remove the speaker grille. Use short, ¼-inch spintite.

CRT Removal

The CRT is removed from the front of all models. It is not necessary to remove the chassis on any model to replace the CRT. Disconnect all appropriate cables (i.e., high-voltage lead, degausser coils). Remove all assemblies from the CRT neck (deflection and convergence yokes, etc.). Remove four CRT mounting screws at the front of the cabinet.

Note: Remove the top two screws first and install the bottom two screws first when installing replacement CRT.

Escutcheon Removal

Remove all knobs, including slide control knobs, by pulling them straight out. Disconnect the speaker after sliding the chassis forward. Remove two screws from the top rear of the escutcheon and two screws from the bottom rear of the escutcheon (see Figure 12–1).

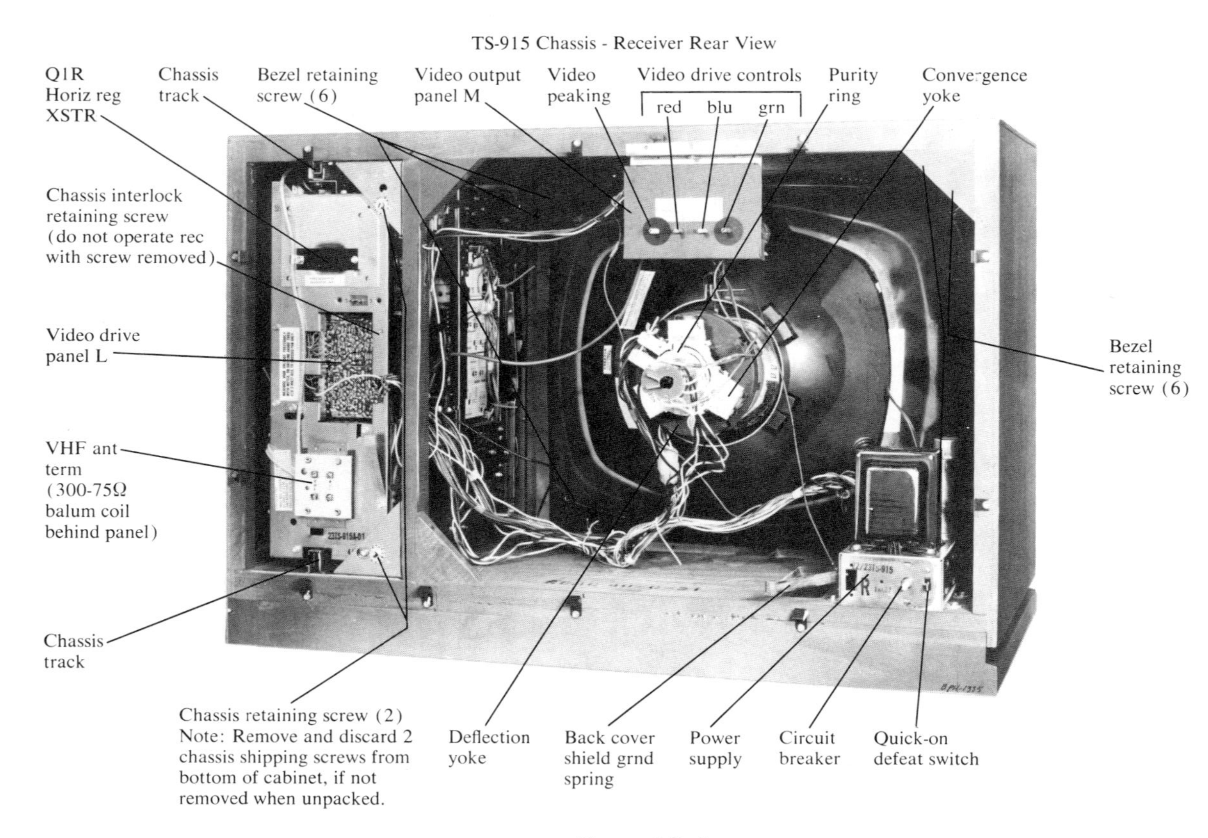

Figure 12–2
Courtesy of Motorola, Inc.

Note: Do not operate the receiver with the speaker disconnected. The audio-output components could be damaged.

Tuner Removal

Vhf-tuner Removal Slide the chassis forward and open the convergence door to its service position. Disconnect all tuner wiring, and identify to facilitate rewiring. Remove the shield at the rear of the tuner and three tuner mounting screws. Remove the tuner from the chassis.

Note: The shield at the rear of the tuner may be soldered in place on some sets.

Uhf-tuner and Drive-gear Adjustment (Pushbutton uhf Versions Only) To insure good repeatability and minimum backlash, the following precautions should be observed when one replaces the uhf tuner:

1. To obtain proper mesh between uhf tuner gear and pushbutton-assembly rocker gear, the mesh is adjustable with the two recessed hex screws located on the front of the control panel (see Figure 12–1 for location). Loosen the two recessed screws. Apply pressure between uhf-tuner body and pushbutton assembly until the gears mesh. Tighten the two recessed screws. Check for station repeatability. Note: If the gear mesh is too loose, excessive backlash will result. If the gear mesh is too tight, poor repeatability may result. Adjust as necessary.
2. To obtain proper mesh between the worm gear and the pushbutton-assembly clutch gear, adjust the set screw on top of the pushbutton assembly for correct mesh. If the mesh is too loose, excessive backlash will result.
3. To reduce worm-gear end play, adjust the screw and nut at the end of the worm gear. If too loose, excessive backlash will result.
4. For clutch adjustment, depress one of the buttons. Note if the clutch gear is disengaged from clutch plate. Release the button. Note if the roller clutch release is free. If necessary, adjust the clutch plate. Loosen the set screw on the clutch plate. Press the clutch plate onto the clutch gear until the clutch release roller is free. Tighten set screw; check for backlash.

Uhf-tuner Removal Slide the chassis forward (Figure 12–1) and open the convergence door to its service position (Figure 12–3). Remove the

TS-915 Chassis - Convergence Panel Opened in Service Position (Video Output Panel Mounted in Service Position)

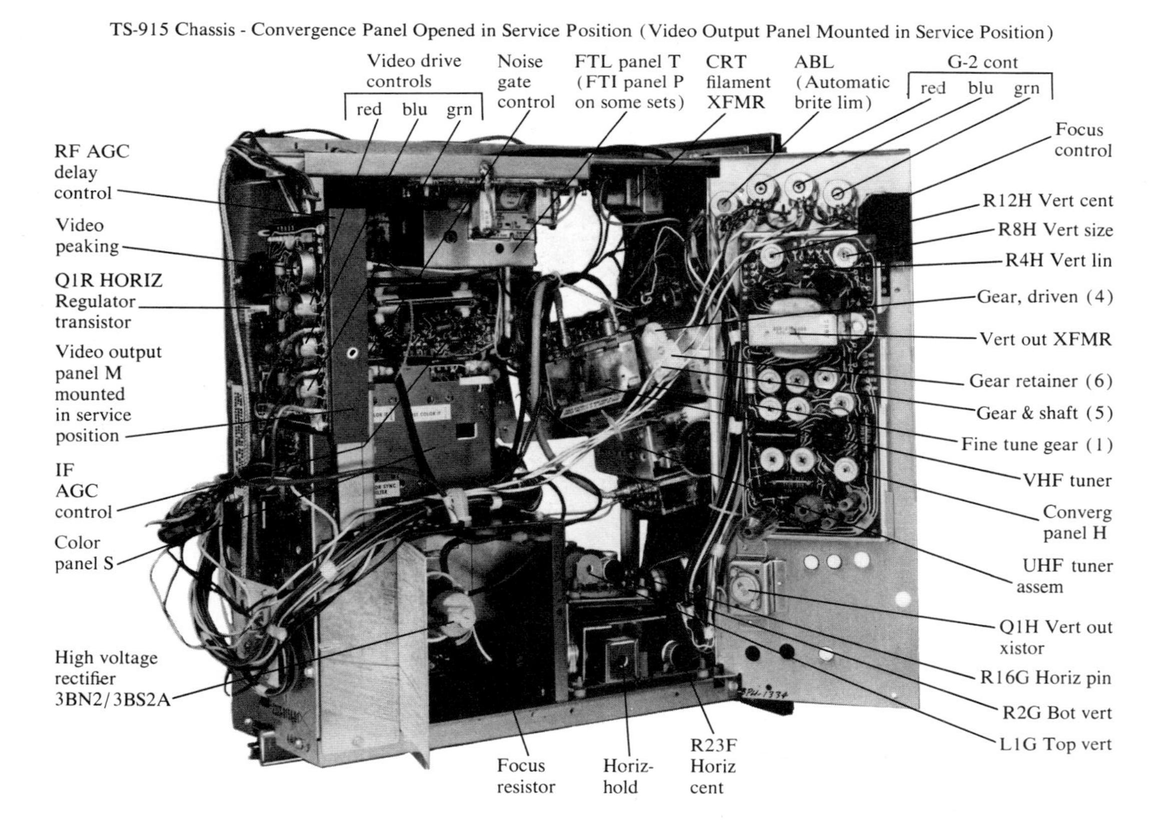

Figure 12–3
Courtesy of Motorola, Inc.

front escutcheon. Disconnect and identify all uhf-tuner wiring. Remove two recessed ¼-inch hex screws at the front of tuner and two $^{3}/_{16}$-inch hex screws at rear of pushbutton assembly. Remove the tuner and separate from the bracket if necessary. To install, set pointer to the low end of the scale and fully close the tuning capacitor in the uhf tuner. This insures proper meshing of the gears and tracking of the dial indicator. Reassemble in reverse order of disassembly instructions. Dimples in the tuner mounting bracket will locate tuner in its proper position.

Uhf-tuner and Pushbutton-assembly Removal The uhf tuner and pushbutton assembly may be removed as one unit. Position the dial pointer at midscale, then remove the scale. Remove four screws at the corners of the pushbutton cutout and two screws at the uhf manual-tuning shaft. Do not remove the two recessed screws.

High-voltage Cage Removal

To facilitate necessary repairs inside the high-voltage cage, the cage can be removed from the chassis as a unit. Remove two screws at top forward end of the cage. Unplug the CRT filament transformer (*T*2J, Figure 12–3 and foldout schematic) mounted on top of the high-voltage cage. Remove all necessary plastic wire ties and loosen two ¼-inch screws that secure the vhf-uhf antenna terminal strip. This allows the high voltage to be tilted toward the rear of the set. Unplug the connectors located on the back of the cage and return the cage to its operating position. Open the access door to the high-voltage rectifier tube and remove two plastic hinge pins to facilitate removal. On some chassis, a two-piece plastic hinge is used. Use a dial screwdriver to pry out the centerpiece before attempting to remove the hinge.

questions

1. Fill in the blank spaces below corresponding to the customer adjustments and indicator lights shown at the right.

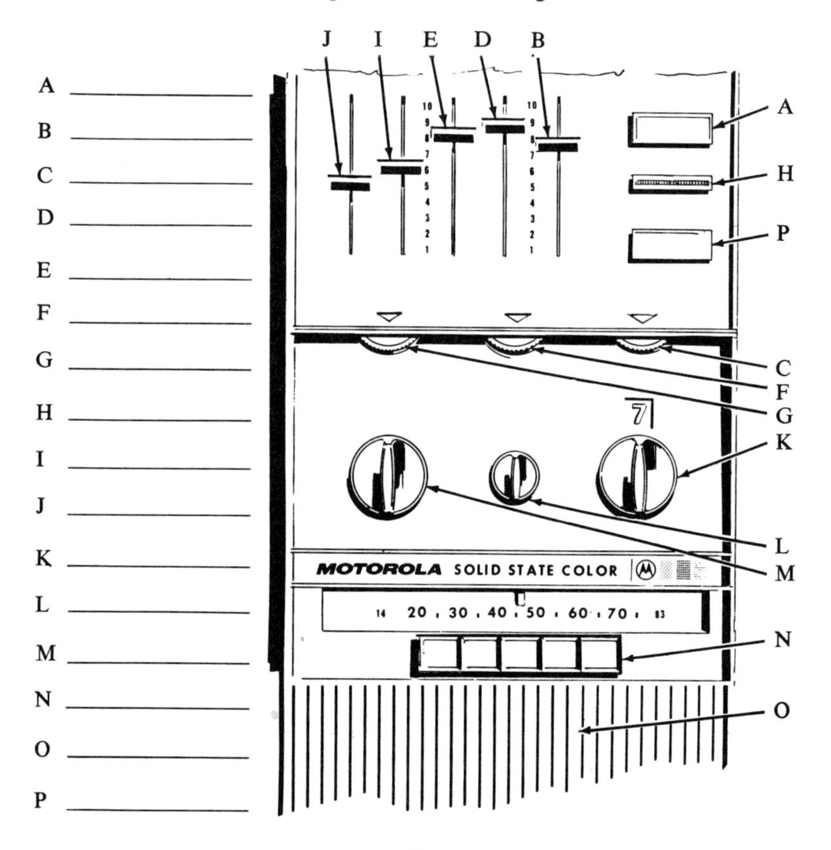

Figure 12–4

2. Describe how to slide the chassis forward from the cabinet for service and operation. Also, how to remove the chassis from the cabinet.
3. Why is silicon grease used on the power transistors described?

13 functional diagram

A working acquaintance with the overall functional diagram is essential to the understanding of the receiver as a system (Figure 13–1). Working knowledge of the system enables a technician to visualize the primary and secondary signal flow, control voltage paths, feedback loops, and operation sequences. Coupled with a good foundation in fundamentals and the ability to use and interpret test equipment, the functional diagram (Figure 13–1) is a valuable piece of test equipment.

The color CRT is actually an oscilloscope which defines defects. Realizing this enables the professional servicer to identify any malfunctioning circuit.

The color CRT is the ultimate display device. The picture means everything to the customer. Failure in almost any circuit will affect the picture. The CRT's requirements, therefore, are a logical starting point for the functional diagram.

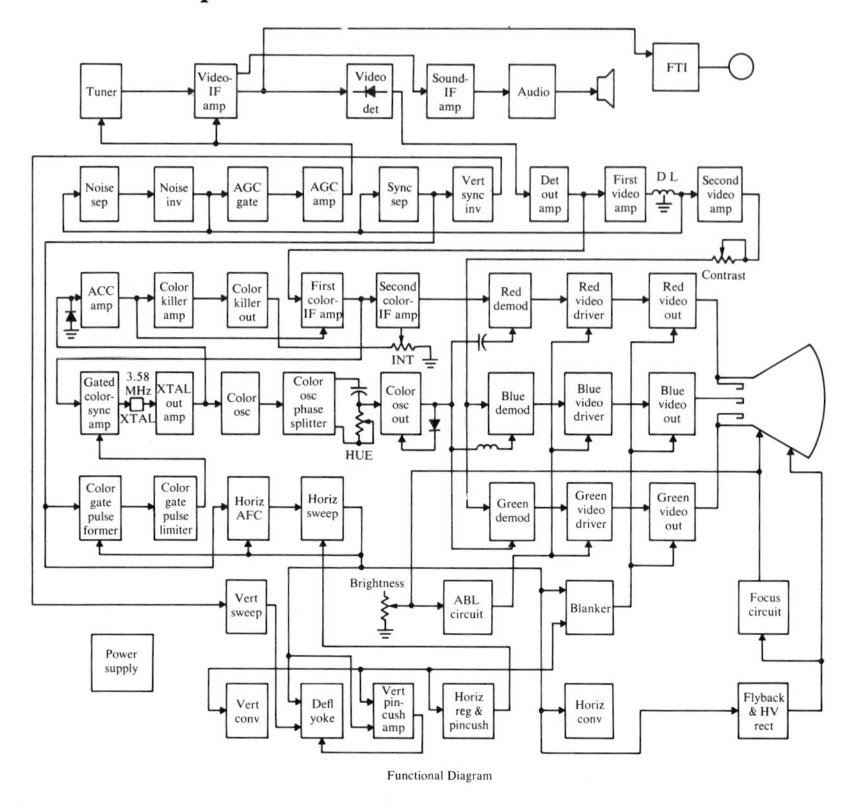

Figure 13–1

CRT Requirements

The red, blue, and green gun assemblies each require approximately 140–165 volts peak-to-peak video signal at their respective cathodes to establish beam blanking at high positive voltage representing vertical and horizontal blanking pedestals. Increased beam current is achieved with lower values of positive video voltages. With 1 to 3 V p-p video developed across the video-detector load, considerable amplification is needed to modulate the three beams from cutoff to maximum. This amplification is accomplished by three sets of video drivers and output amplifiers operating in parallel. Each set drives its own related electron gun at the CRT. In common with the three pairs of drivers and amplifiers are two stages of video amplifiers with a delay line inserted between them. There are no video amplifiers in this receiver in the usual sense. Rather, there are three pairs of video drivers and output amplifiers which amplify the conventional monochrome composite-video signal during black-and white-reception. With color reception, the three pairs become color-video amplifiers. The red video driver and the output pair supply the "red" cathode of the CRT with video signals. The blue pair supplies the "blue" cathode and the green pair drives the "green" cathode. Color signals are narrow-band compared to the total information required to reproduce a picture. Video-modulating voltages beyond the color-frequency response are transmitted monochrome. Therefore, the three sets of video drivers and amplifiers are color-video amplifiers only for those portions of a picture containing color. For those portions of the reproduced scene containing only brightness information, the three sets of drivers and amplifiers are truly conventional video amplifiers. It may be recalled from the study of basic color systems that fine detail is not transmitted in color because the human eye is essentially blind to it (especially to those colors in the blue region of the spectrum). The eye is somewhat less color-blind to those colors that range from orange to cyan.

There are no color difference-signal amplifiers ($R - Y$, $B - Y$, $G - Y$) in this receiver. No matrixing of the brightness signal (Y) with color difference signals ($R - Y + Y = R$) is required at the CRT to produce beam modulation with color signals. Instead, simultaneous demodulation of the composite color signal, which includes the brightness signal, occurs in each of the three demodulators.

The output of each demodulator is a true color signal during color reception. During black-and-white reception and for those portions of a color program transmitted black-and-white, composite-video information passes through each of the three demodulators in parallel without effect because the demodulators are a virtual dead short for this signal. It can be stated that each of three electron guns in the CRT has its own related demodulator, video driver, and video-output amplifier stage. For black-and-white reproduction, each configuration accommodates the same signal, with only slight differences in amplitude, to achieve proper video gray-scale tracking. For color reproduction, the output of each video configuration is different, depending upon the color demodulated.

Beam Blanking

It is desirable to cut off the three electron beams during horizontal and vertical retrace to prevent retrace lines, which contain no camera signal, from appearing superimposed on the reproduced picture. Normally, blanking voltages, which are part of the transmitted video signal, perform this function. A supplementary beam-blanking system is desirable to compensate for incorrect brightness and contrast adjustments by the set operator. The customer might set contrast and brightness controls in such a manner that blanking voltages cannot perform the purpose intended. Because the three video-output stages are direct coupled to each related CRT cathode, a change in conduction affects beam current. Therefore, it is convenient to cut off the three output stages to blank off each beam. The emitter of each video-output transistor finds its way to ground through the saturated half of a solid-state switch. When the switch is opened, the video emitters are removed from ground, cutting these stages off. Collector voltages jump up to source value along with the direct-coupled CRT cathodes, cutting off beam current. The switch is opened by application of pulses, derived from the horizontal and vertical circuits, to the input transistor of the switch. Conduction of the switch of the input transistor as a function of these pulses opens the normally saturated switch output transistor, cutting off CRT beam current as described. This supplementary beam-blanking circuit is shown on the functional diagram as the "blanker."

Automatic Beam Limiter

Another CRT requirement is limitation of maximum beam current to a value that can be delivered safely by the horizontal-deflection flyback power supply (high-voltage supply). This is accomplished by controlling the conduction of the video drivers with an automatic-brightness-limiter circuit. With direct coupling between the video-driver and video-output stages, output conduction is controlled. As previously explained, direct coupling between the output stages and the CRT cathodes permits beam control. The automatic-brightness-limiter configuration is a two-stage dc amplifier which senses a change in beam current by monitoring focus voltage bled from a divider. Focus voltage will drop with an increase in beam current. A drop in focus voltage results in a decrease in conduction of the first dc amplifier in the limiter circuit. Connected as an emitter follower, the emitter voltage drops with focus voltage. By direct coupling, the second dc amplifier decreases conduction. Also an emitter follower, its emitter voltage drops, causing reduced forward bias at the video drivers and resulting in a decrease in conduction. With increased impedance between the driver collector and the 35-V dc supply, there is less forward bias in the following direct-coupled (no coupling capacitor) video-output transistor's base. Its conduction will decrease, accompanied by the rise in collector voltage toward the 255-V dc source (which biases the CRT cathodes toward cutoff), reducing beam current. An additional advantage of beam-current control is to provide balanced relationships between contrast and brightness. With direct coupling of the composite-video signal from the detector to the CRT, changes in contrast-control settings will change the dc bias levels at the CRT, shifting brightness. The automatic-limiter circuit affords independent control of contrast to compensate for changes in brightness caused by the contrast-control adjustment.

High Voltage and Focus

The required 27 kV for the CRT second anode is developed by a straightforward high-voltage supply. The flyback system, driven by the horizontal-deflection circuitry, is conventional, employing the only vacuum tube in the receiver (a 3BN2 high-voltage rectifier). Focus voltage is established with a resistive bleeder network. Distributing the focus voltage via variable resistors (potentiometers) permits G-2 setup for gray-scale background. An additional potentiometer permits customer adjustment of background to satisfy individual preferences. This control is labeled "Tint."

The Video-IF Amplifier

Ahead of the two video amplifiers is a three-stage video-IF amplifier suitably trapped to prevent response to undesired upper and lower adjacent channel frequencies. The traps are the major response determinants because interstage coupling is broad.

A forward AGC system holds the IF amplifier gain constant despite changes in antenna signal. An automatically variable forward bias is applied to the base of the second IF amplifier. The gain of the first IF amplifier is controlled by the emitter voltage of the second. This technique permits gain control of two stages by direct control of one, and eliminates the possibility of cross modulation of various signals at the first common IF-amplifier stage. Cross modulation is less likely to occur later in the IF amplifier because of stringent trapping earlier.

The Fine-Tuning Indicator

The fine-tuning indicator is an exciting customer feature which solves the problem of incorrect fine tuning. When the receiver is correctly tuned, a voltage-sensitive circuit shorts an indicator lamp connected from 55 V dc to ground. Under these circumstances, the tuner's local oscillator heterodynes (beats) with the selected channel picture carrier to produce a picture IF of 45.75 MHz. A high *Q* circuit, resonant at this frequency, develops maximum voltage across it. When rectified, this voltage triggers the circuit. Incorrect fine tuning will produce an incorrect picture-carrier frequency. The circuit will not react, and the indicator lamp will remain lit to warn the customer that the receiver is not adjusted correctly.

The Tuner

The tuner is a three-stage subassembly employing an RF amplifier, mixer, and local oscillator. Delayed forward AGC bias is used to vary the RF gain. A four-circuit wafer-switch assembly assures excellent selectivity and bandwidth. Noise figures are somewhat better than in vacuum-tube tuners.

The Noise Inverter

Composite video, which may contain noise spikes, is taken from the input of the second video amplifier and coupled to a noise separator which is biased to cut off at sync-pulse amplitude. Noise pulses in excess of sync-pulse amplitude pass through the noise separator, are amplified, inverted, and placed back on the composite signal in opposite phase to cancel the original noise. Noise cancellation renders the deflection circuits immune to inadvertent triggering which could cause horizontal tearing and vertical rolling.

The Sound System

Sound (41.25 MHz) is taken from the third video-IF amplifier and applied to a transistor detector which heterodynes (beats) the picture carrier (45.75 MHz) against the sound carrier to produce the 4.5-MHz FM sound carrier difference frequency. An integrated circuit follows the 4.5-MHz beat detector to perform the functions of amplifying the 4.5-MHz beat and detecting the audio. A conventional 8-ohm loudspeaker is driven by a three-stage audio amplifier.

Other Composite-Video Paths

In addition to going to the three demodulators in parallel, composite video is applied to the input of an AGC gate which is keyed on by a pulse supplied from horizontal-deflection circuitry. Sync-pulse amplitude is proportional to received carrier strength and provides an excellent reference. Since horizontal-sync pulses and keying pulses coincide in time, the amount of conduction by the gate is proportional to sync-pulse amplitude. The collector current thus obtained determines the amount of conduction of a dc amplifier called an *AGC amplifier*. This second stage supplies the actual forward AGC bias to the base of the second common IF transistor. The same voltage is delayed by a reverse-biased diode for application to the RF stage of the tuner. Both AGC threshold and RF delay are adjustable.

Composite video is also coupled from the second video amplifier to a sync separator where horizontal-sync and vertical-sync pulses are stripped from the total signal in a typical signal-biased arrangement. Separated sync is coupled to the horizontal AGC network for control of horizontal-deflection frequency. Also, a vertical-sync inverter turns vertical sync around (from negative-going to positive-going) suitable for triggering the vertical-deflection generator.

Other CRT Requirements

In addition to 140-V to 165-V p-p video, 27-kV second-anode voltage, focus voltage, beam blanking, and beam-current limiting, the CRT requires horizontal and vertical deflection of its three electron beams. Further, the rasters produced by such compound deflection must be corrected for the inherent size and shape distortion, commonly called pincushioning, which occurs on flat-surfaced viewing screens. Finally, registration of the three rasters must be established by horizontal-convergence and vertical-convergence circuits for each of the three beams.

Horizontal Deflection

A single-stage oscillator generates the horizontal-deflection signal which is amplified and passed on to switch a horizontal-driver transistor on and off. This stage, in turn, switches a parallel pair of horizontal-output transistors off and on. A diode damper rectifies shock-excited oscillations. The rectified power is used for beam deflection from start of horizontal scan to screen center, after which the output pair conducts to deflect the beams from screen center to end of trace, completing the full line scan. At this instant retrace (flyback) is initiated. A pulse-limiting diode saves the output pair from breaking down as a result of a high, rapid rise in collector voltage which occurs when the pair is cut off. Horizontal B+ is modulated by a 60-Hz parabola taken from the vertical circuit. This technique enables horizontal width to be increased during each field (60 fields a second) to correct horizontal pincushion distortion. At the same time, the average horizontal B+ is regulated to 80 V dc at zero beam current, and to about 76 V dc at maximum beam current.

Horizontal signals are modified by a convergence circuit which supplies a parabola of current for each of the three convergence coils.

Vertical Deflection

A free-running solid-state switch supplies the deflection yoke with a 60-Hz current sawtooth which deflects the three beams vertically. Additional stages provide wave-shaping and current-limiting functions. Vertical pincushion distortion is corrected by supplying the vertical deflection yoke with a horizontal signal which deflects each horizontal line vertically. Therefore, vertical deflection is corrected at a horizontal rate to make the top and bottom of the raster straight. Convergence signals are obtained from both the vertical deflection yoke and vertical-output circuit.

The Color Selection

The 3.58-MHz color signal, available as part of the composite-video signal across the detector load, is passed through two stages of color-IF amplification after coupling through an emitter follower. The total band-pass charateristic is symmetrical and centered at about 3.58 MHz. The input to the first color-IF amplifier is tuned to 4.1 MHz to compensate for the reduced gain of these sidebands, due to the response characteristics of the video-IF amplifier.

The gain of the first color-IF amplifier is held constant by an automatic color-control circuit, while the second color-IF amplifier is switched on or off by a color-killer circuit. The switching voltage is made variable manually as a means of controlling color intensity (saturation) by the customer. The amplified color-signal output from the color-IF section is coupled to the three demodulators where it is combined with an appropriately phased 3.58-MHz signal and the brightness signal to produce color-video voltages.

These voltages are then presented to the related video driver and output transistors for beam modulation of the appropriate electron beam.

The 3.58-MHz CW Reference Signal

Demodulation can occur only when there is a carrier from which the information-carrying sidebands can be extracted in the form of useful voltages. Because the original carrier is suppressed at the transmitter, it must be reinserted at the receiver. This is accomplished by a local color oscillator often called a CW reference oscillator. The important requirement for the CW reference signal is that it must be synchronized in frequency and phase with the original suppressed carrier, a sample of which is actually transmitted as a form of color sync. Color sync is transmitted during a short interval coincident with horizontal blanking, immediately after horizontal sync (the back porch). In this way, no color-sync information can appear on the screen. Also, the CW reference signal applied to the demodulators is altered in phase to establish demodulation axis. This receiver employs unique CW reference phase angles of 76.5° from color sync for the red demodulator, 193° from color sync for the blue, and 120° from color sync for the green. The latter angle provides maximum output at what might appear to be the magenta quadrant of the color vector diagram, but a reversal of the diodes in the green demodulator, with respect to the red and blue, enables maximum output 180° out of phase with magenta, which is green.

The color oscillator is a single-stage transistor whose frequency determinants are carefully designed *L/C* circuits and a 3.58-MHz CW synchronizing signal which is generated by a ringing crystal. The crystal is caused to vibrate at its natural frequency when shocked by color sync. These vibrations cause the oscillator to synchronize in frequency and phase, assuming, of course, that the *L/C* circuits are tuned to the same frequency.

The output frequency of the oscillator is coupled to a phase splitter which produces two oscillator voltages 180° out of phase with respect to each other. Therefore, one of these voltages is in phase with color sync, the other is 180° out of phase. An *R/C* circuit, between the phase splitter and the following 3.58-MHz amplifier, establishes a phase angle which can be varied by the customer to establish proper flesh tones as a color reference in the reproduced picture. This adjustment is a variable risistance called a "hue control." The phase angle is continuously variable over a range of 140°, permitting wide latitude as compensation for transmission differences, antenna characteristics, component aging, video tape, and other factors which might shift color-sync phase or color-signal phase. The 3.58-MHz color oscillator output amplifier provides isolation between the hue-shifting configuration and the demodulators. It also affords a measure of protection against changes in CW-reference-signal amplitude applied to the demodulators. A fixed phase-shifting network between the 3.58-MHz output amplifier and the demodulators establishes phase angles for each of the three demodulators consistent with demodulation along the red, blue, and green axis as described.

Color Sync

The full composite color signal appears at the input of the gated color sync amplifier. All of this signal must be rejected at this point except for the eight or nine cycles of 3.58-MHz color sync which occurs during horizontal blanking, after horizontal sync. The gated color-sync amplifier is turned on by a gating pulse which occurs during the described blanking interval. When the gate is opened by the gating pulse, color sync passes through the open gate to ring the crystal. When the gating pulse passes, the gate slams shut to prevent the passage of any other information.

Automatic Color Control

To allow a wider range in fine tuning for color and to hold color saturation constant despite changes in amplitude, a one-stage dc amplifier is employed to automatically vary the gain of the first color-IF amplifier transistor.

When color sync is present, indicating a received color transmission, the output of the crystal amplifier is proportional to color-signal amplitude. The output voltage is rectified, creating a dc control voltage of positive polarity which is applied to the base of the AGC amplifier and causing increased conduction. The extent to which the AGC amplifier conducts is proportional to color-sync amplitude. Forward bias for the first color-IF amplifier is obtained from the 35-V dc supply, with the AGC amplifier acting as a variable impedance to ground on the bias supply line. Increased conduction of the AGC amplifier, because of a strong color signal, causes the forward bias of the first color-IF amplifier to be reduced, resulting in less color amplification. A decrease in color-signal amplitude results in increased forward bias at the first color-IF amplifier and a greater color amplification signal. To assure maximum weak signal gain, AGC voltage is delayed by a reverse-biased diode before application to the first color-IF amplifier.

Conduction of the AGC amplifier as a function of received color sync removes the reverse bias at the base of the first color killer. This allows the color-killer solid-state switch to turn on.

The Color Killer

To prevent passage of miscellaneous video signals and noise through the color-IF section when no color program is received, a two-stage voltage-sensitive solid-state switch allows the second color-IF amplifier to drop into cutoff. When color sync is present, the AGC amplifier turns the color-killer switch on. Conduction of the color-killer amplifier (a PNP transistor) switches on the color-killer output (an NPN stage) by direct coupling. The voltage developed by the color-killer output forward biases the second color-IF amplifier to bring it out of cutoff, allowing passage of color signal. The forward bias is made variable manually as a customer adjustment to set color intensity (saturation).

The Demodulators

Dual diode demodulators are employed in a special configuration that resembles a phase detector in certain respects. When color and CW reference signals are present, demodulation of color sidebands occurs to produce color-video voltages. When color and reference signals are in phase, maximum negative output voltage is produced. When 180° phase relationship occurs, maximum positive output voltage is developed. When the two are in quadrature (90°), zero output occurs because the CW reference signal is going through zero. Absence of reference signal prevents demodulation.

CW reference phase is fixed, while color-signal phase constantly changes depending upon what hue is transmitted at any given instant. The unique feature of this system is that three demodulators provide the three color signals without having to matrix (add) portions of the blue and red signals to obtain green. This is different from most previous receivers in the industry. Also, the presence of the composite color and brightness signals in the demodulators directly produces color-video voltages. This eliminates the need for color-difference amplifiers (*R-Y, B-Y,* and *G-Y* amplifiers) and also allows the advantage of driving only one element (the cathode) of each electron gun, rather than two elements (cathode and grid). This eliminates gamma distortion and nonlinear matrixing of the brightness signal with the color-difference signal at the CRT.

The Power Supply

Four silicon diodes in a full wave bridge supply high current which is well filtered by a choke-input-type network. High current is demanded by several sections of the receiver including audio output, horizontal deflection, vertical output, etc. Two diodes of the bridge serve as a full-wave rectifier to supply 35 V dc for most of the primary signal circuits of the receiver. A half-wave rectifier develops 255 V dc for the video-output amplifiers. Various other voltages are divided off the main bridge supply for specific needs throughout the receiver. Automatic degaussing is featured with a thermal relay to bypass the degaussing coil when the receiver is operating. CRT filaments are always energized, whether the set is on or off, to permit longer filament life and a quick warm-up. A defeat switch is provided to de-energize the CRT filaments if desired. The receiver is protected against high current drain by a conventional thermal cutout.

questions

1. What type of AGC is used in the tuner?
2. Is forward or reverse AGC used in the video-IF amplifier?
3. What are the frequencies of the following signals in the video-IF amplifier?
   a) IF-picture carrier
   b) IF-sound carrier
   c) IF-color subcarrier
4. A sound-beat carrier of 4.5 MHz is created in what section?
5. What function does the IC perform?
6. The color signal is coupled to the first color-IF amplifier from what stage?
7. What frequencies are involved in the color-IF amplifiers?
8. Is the 3.58-MHz crystal required to assure that the color oscillator operates?
9. What functions do the demodulators perform during a black-and white-program?
10. What three signals are directly demodulated by the demodulators?
11. What polarity video signal is required to turn the CRT off?
12. Why do we need the color-killer circuit?
13. The color reference signal is:
    a) 4.5 MHz
    b) 41.5 MHz
    c) 3.58 MHz
    d) 45.5 MHz
14. Why is this reference signal needed?
15. The heterodyne of 41.25 MHz with 45.75 MHz produces what frequency?
16. Draw a block diagram of a color TV receiver and indicate signal flow.

14 in-home service

Introduction

This chapter should provide an orderly, logical plan for service analysis of solid-state color-television receivers. Receiver operation can usually be restored through the isolation of the defect to a panel and replacement of it. Since most panels contain partial or complete circuit functions, you must follow a logical procedure for determining which panel is defective. Consider, for example, the vertical-sweep circuits. Three of the vertical transistors are located on the video-amplifier panel, one is on the convergence panel, and one is mounted on the main chassis.

We can isolate a defective panel effectively with very little knowledge of receiver circuitry. The only instrument required is an accurate vacuum tube voltmeter (VTVM) or a 20,000 ohm per volt voltmeter (VOM).

Each panel has a letter assigned to it. All components and terminal connections carry this panel letter as a suffix after every reference number. For example, the video-IF is the "B" panel. All reference numbers relating to components on that panel carry a suffix "B" (*R*–10B, *C*–10B, etc.).

A panel block diagram is shown in Figure 14–1. The blocks outlined in heavy print represent the individual replaceable panels. Circuits on these panels are marked accordingly. The blocks outlined in light print represent circuits not located on a replaceable panel. These circuits are on the main chassis. Figure 14–1 is a block diagram of the receiver and describes the function of the panel plus its suffix letter.

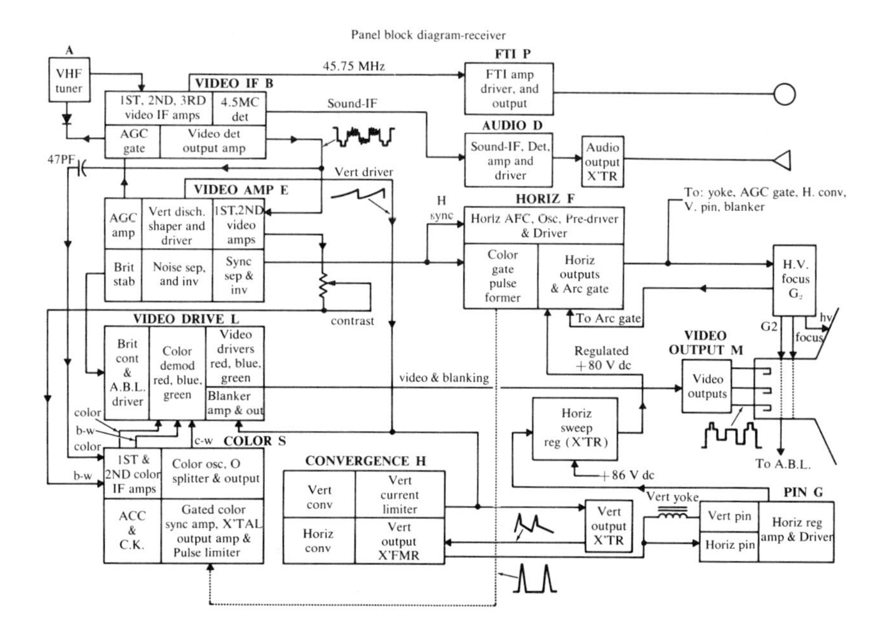

Figure 14–1

Figures 14–2 and 14–3 are views of the top chassis layout. They locate the physical layout of the panels and the test points of each. The chassis components not mounted on panels are also located for the serviceman.

Much time can be saved by checking the following before proceeding any service job:

1. Give the receiver a visual check.
2. Make sure all panels are secured to the chassis mounting studs.
3. Inspect and make certain all connecting plugs are tight.
4. Check to be certain controls are set properly.

Previously we determined that the color receiver was composed of two signal paths, the black-and-white and the color signal paths. Black-and-white troubles fall into four categories: general receiver troubles, raster troubles, sound troubles, and picture troubles.

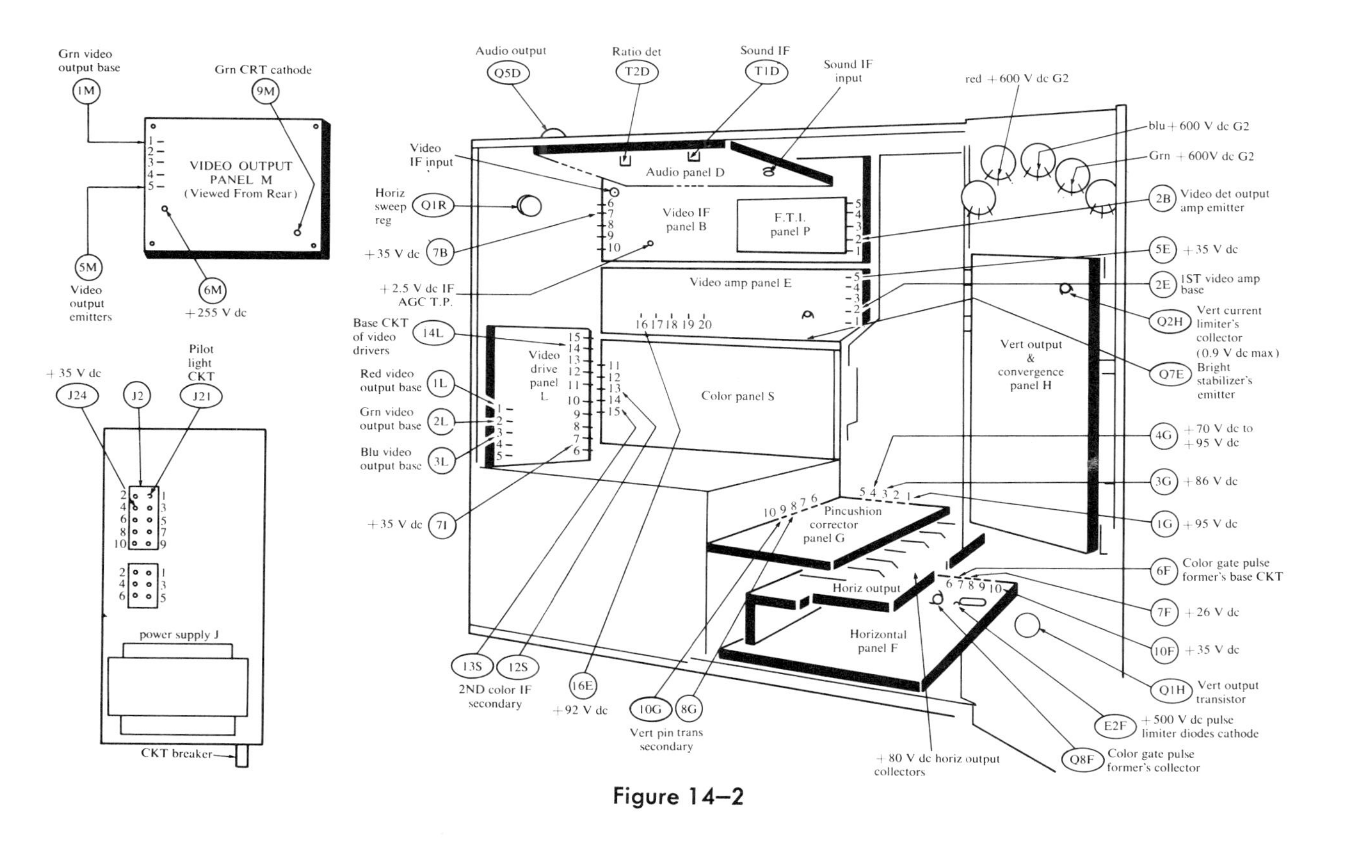

Grn video output base
1M
Grn CRT cathode
9M
VIDEO OUTPUT PANEL M (Viewed From Rear)
5M
Video output emitters
6M
+255 V dc
+35 V dc
J24
J2
Pilot light CKT
J21
power supply J
CKT breaker
Audio output
Q5D
Ratio det
T2D
Sound IF
T1D
Sound IF input
red +600 V dc G2
blu +600 V dc G2
Grn +600V dc G2
Video IF input
Audio panel D
Horiz sweep reg
Q1R
Video IF panel B
F.T.I. panel P
2B
Video det output amp emitter
+35 V dc
7B
5E
+35 V dc
2E
1ST video amp base
+2.5 V dc IF AGC T.P.
Video amp panel E
Q2H
Vert current limiter's collector (0.9 V dc max)
Base CKT of video drivers
14L
Video drive panel L
Vert output & convergence panel H
Q7E
Bright stabilizer's emitter
Red video output base
1L
Color panel S
Grn video output base
2L
4G
+70 V dc to +95 V dc
Blu video output base
3L
3G
+86 V dc
+35 V dc
7I
Pincushion corrector panel G
1G
+95 V dc
6F
Color gate pulse former's base CKT
Horiz output
7F
+26 V dc
Horizontal panel F
10F
+35 V dc
13S
12S
2ND color IF secondary
16E
+92 V dc
10G
8G
Vert pin trans secondary
Q1H
Vert output transistor
E2F
+500 V dc pulse limiter diodes cathode
+80 V dc horiz output collectors
Q8F
Color gate pulse former's collector

Figure 14–2

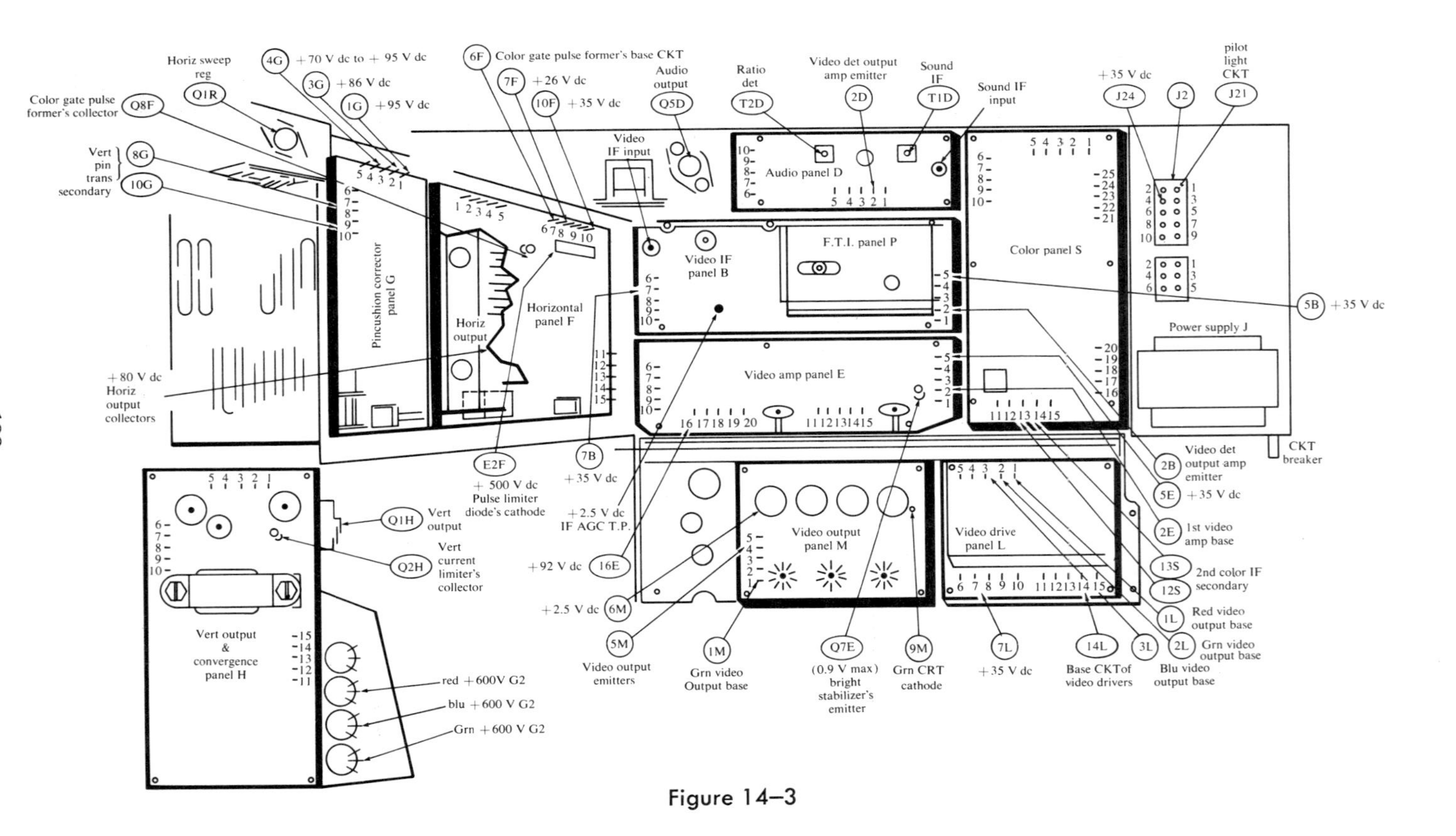

Horiz sweep reg
Q1R
4G +70 V dc to +95 V dc
3G +86 V dc
1G +95 V dc
6F Color gate pulse former's base CKT
7F +26 V dc
10F +35 V dc
Audio output
Q5D
Ratio det
T2D
Video det output amp emitter
2D
Sound IF
T1D
Sound IF input
+35 V dc
J24
J2
pilot light CKT
J21
Color gate pulse former's collector
Q8F
Vert pin trans secondary
8G
10G
Video IF input
Audio panel D
Color panel S
F.T.I. panel P
Video IF panel B
Pincushion corrector panel G
Horizontal panel F
Horiz output
5B +35 V dc
Power supply J
Video amp panel E
+80 V dc Horiz output collectors
CKT breaker
7B
+35 V dc
E2F
+500 V dc Pulse limiter diode's cathode
+2.5 V dc IF AGC T.P.
2B Video det output amp emitter
5E +35 V dc
Q1H Vert output
Q2H Vert current limiter's collector
+92 V dc
16E
Video output panel M
Video drive panel L
2E 1st video amp base
13S
12S 2nd color IF secondary
+2.5 V dc
6M
1L Red video output base
Vert output & convergence panel H
5M
Video output emitters
1M
Grn video Output base
Q7E
(0.9 V max) bright stabilizer's emitter
9M
Grn CRT cathode
7L
+35 V dc
14L
Base CKTof video drivers
3L
Blu video output base
2L Grn video output base
red +600V G2
blu +600 V G2
Grn +600 V G2

Figure 14–3

General Receiver Troubles

1. Completely dead
2. No sound, no picture, but a normal raster (picture tube lit-up across the face of the tube both in the vertical and horizontal direction)
3. No raster, no noise from speaker, but lit dial lights

Raster Troubles

1. No brightness
2. Inadequate height
3. Poor focus
4. Little or no vertical sweep

Picture Troubles

1. No vertical sync, normal horizontal sync
2. No horizontal sync
3. No horizontal or vertical sync
4. Improper gray scale (poor black-and-white picture)

Sound Troubles

1. No sound, picture normal
2. Distorted sound (hum, buzz, weak)

Color problems fall into three categories: no color, no color sync, and wrong color.

The wrong-color defect is broken down into two characteristics—primary color absence and primary color predominance.

These system problems are common to all makes of television receivers. The logical approach to analyzing them is also held in common, although the specific servicing procedure used will vary because of layout and circuit differences. The following method of analysis may be called the "Talk-to-the-receiver-method." Using this method, one asks the right questions to get the right answers. The method should lead to the accurate isolation of the cause of the problem and to its proper correction. We will use it on the black-and-white system problems first.

Completely Dead Receiver

Q.1. Turn the receiver on and depress the breaker momentarily. Does the dial light turn on and remain on?

A.1.N. No? Then disconnect the power-supply plug, (J2, Figure 14–2). Connect the jumper between the brown lead on the open plug and vacated receptacle (J2–1, Figure 14–2). The brown lead connects to the dial light and J2–1 is the output of the power-transformer secondary winding used to supply power for the dial-light bulbs. Depress the circuit breaker. Now does the light turn on and remain on? Still no? Then check the following: dial lights, line cords, power-supply plugs, degaussing-coil plugs, circuit breaker. If these are all right, then the defect is in the power supply. Replacing the power supply should make the set operable.

A.1.Y. If the pilot light does turn on and stay on, we can assume that the power supply is all right. We should now reconnect the power-supply plug (J2, Figure 14–2) and proceed to remove one panel at a time, turning the receiver on after each removal. When the defective panel has been removed, the circuit breaker will remain set and the dial light will remain on.

No Raster, No Noise, Dial Light On

Q.1. If the dial light is on and remains so, check to see if there is a raster and sound. If there is no raster and no sound, measure the voltage at the red lead on the large power-supply plug. Does the voltage measure at least 35 V dc?

A.1.N. No? Then disconnect the large power-supply plug and measure the voltage at receptacle J2–4 (Figure 14–2). If the voltage still does not measure at least 35 V dc, we can conclude that the power supply is defective.

A.1.Y. If voltage does measure at least 35 V dc, remove the ac power plug. Connect an ohmmeter (R × 100 scale) between the red lead on the disconnected plug and chassis. Remove the panels, one at a time, while observing the ohmmeter. The panel that causes a significant change in the ohmmeter reading is defective. Normally the resistance measured should remain fairly constant.

To summarize, we can say that a dead receiver can be the result of incorrect or zero voltage from the power supply. The problem can be the result of a defective power supply or external loading. Isolating the supply by disconnecting and testing as described should differentiate between the supply and chassis circuit loading. It is a process of elimination.

No Sound, No Picture, Normal Raster

Q.1. Adjust brightness and contrast to maximum. There should be a raster. Set the channel selector halfway between adjacent channels and measure the voltage at IF. AGC (automatic gain control) is located on panel B (see Figure 14–2). Does it measure 2.5 V dc plus-or-minus 0.3 V dc?

A.1.Y. Yes? Then remove the video-IF cable from the IF panel (B). Touch the cable receptacle on the panel with a metal tool. Does noise appear on screen? If so, it can be concluded that the vhf tuner is defective. Check voltages to the tuner before removing it. If noise does not appear on the screen, it can be concluded that the video-IF panel (B) is defective.

A.1.N. If we do not measure 2.5 V dc at the AGC test point, we are not getting AGC voltages. This has probably caused overloading and there is no sound or picture.

Q.2. Now we must ask another question. Is the voltage at the IF-AGC test point too high?

A.2.Y. Yes? Then remove the video-amplifier panel (E, Figure 14–2). If the voltage at the IF-AGC test point is still too high, we can suspect that the video-IF panel (B) is defective. (This panel contains the AGC amplifier.)

A.2.N. If the voltage is not too high, reinsert the video-amplifier panel (E). Short terminal 2B (Figures 14–2 and 14–3) to the chassis. This is the video-detector output of the amplifier emitter. If the AGC is still high we can conclude that the video-amplifier panel (E) is defective. If the AGC voltage is not high then the IF panel (B) is defective. By shorting-out test point 2B we are eliminating the video panel (E) from the circuit. Therefore, it is not contributing its correct effect to the AGC voltage amplifier. On the other hand, if the voltage does go down, the imbalance is in the AGC-amplifier panel (B). The proper balance between the video amplifier panel (E), and the video-IF amplifier (panel B), will yield the 2.5 volts.

No Raster, No Noise From Speaker

Note: To proceed with this portion of the analysis it is essential to have the following tools: a VTVM or 20,000 ohms per volt (volt ohmist), a 2,200-ohm resistor, a 22,000-ohm resistor, a 10,000-ohm resistor. All the voltages are measured with respect to the chassis unless otherwise indicated. Remember, this holds true for this specific receiver model. Resistors may or may not be needed for other receiver analysis. The service manuals will inform you as to the requirements for the specific receiver under analysis.

Q.1. First advance brightness and contrast to maximum. Bridge the 10,000-ohm resistor from the green CRT cathode (9M, Figure 14–3) to chassis. Does the raster turn green? (Remove resistor after test.)

A.1.N. No? Then, set the green screen control to maximum (clockwise). Is the voltage at the slider terminal (G2 for green, Figure 14–2) approximately 500 V dc as seen on a VTVM or 20,000 ohms per volt voltmeter? If so, it can be concluded that the CRT is defective. If not, check the voltage at the cathode of the pulse-limiter diodes. Is it approximately 500 V dc? This test point is located on the horizontal panel (F, Figure 14–2).

If the voltage is not 500 volts, check the collector voltage of the horizontal-output amplifiers. (See Figure 14–2 for location.) It should read approximately 80 V dc or higher. If it does not, test the voltage at test point 4G (Figure 14–2). This test point is located on the pincushion-corrector panel. The voltage should be between 70 and 95 V dc. If it is not, it can be concluded that the pincushion panel (G, Figure 14–2) is defective or there is a loss of source voltage at test point 1G (95 V dc), 3G (85 V dc), shown in Figure 14–2. The source voltage comes from the power supply.

If the voltage at the cathode of the pulse-limiter diodes reads 500 volts dc, it can be concluded that the high-voltage rectifier, the high-voltage transformer, or the focus resistor is defective.

If the collector voltage of the horizontal-output amplifiers is approximately 80 V dc, either the horizontal panel (F) is defective or we have a loss of the 36-V dc source at point 7F or the 35-V dc source at test point 10F (Figure 14–2). If the voltage at 4G (Figure 14–2) is between 70 and 95 V dc, the defect is in the high-voltage regulation or related components.

Through this path of analysis we have eliminated the problem that can cause lack of raster due to problems in the picture tube, high-voltage-panel functions, high-voltage transformer, focus block, and/or high-voltage rectifier. Loss of source voltage (B supply) to any of the above stages can also cause no raster.

Unfortunately these are not the only problems that result in lack of raster. Let us go back to the first test we made. We advanced the brightness and contrast to maximum and bridged a 10,000-ohm resistor from the green CRT cathode (9M test point) to the chassis. Then we assumed the raster did not turn green. Now we must assume the opposite and proceed with the appropriate analysis.

Q.2. First, if the raster turns green, we can definitely assume we have high-voltage and sweep currents. Return the contrast to minimum. Is there any brightness?

A.2.N. No? Connect a jumper from the common terminal of the video-output emitters (5M test point) to the chassis. (See Figure 14–2 for location.) This will eliminate the blanker circuit. Does brightness return? (Remove jumper after performing this test.) If not, bridge a 10,000-ohm resistor from the collector to the base of the green video-output amplifier, test point 6M to 1M (Figure 14–2). This bypasses the green video output. Does the green raster appear? (Remove the resistor after test.) If not, we can conclude that the defect is in the video panel (M, Figure 14–2). This panel is obviously cutting off the CRT guns.

A.2.Y. If there is brightness with the contrast control at minimum, return it to maximum. Bridge a 10,000-ohm resistor from the collector to the base of the first video (Figure 14–2, test points 2E to 5E). This act forward biases the first video-amplifier transistor. If the brightness returns, we can conclude that the video-IF panel (B) is defective or there is a loss of 35-V dc source at test point 7B (Figure 14–2). Panel B was the forward-bias source. The 10,000-ohm bridge replaced this panel. Therefore our conclusion is correct based on the result of the returning brightness.

If the brightness does not return, the video amplifier (E) is defective or we have a loss of 35-V dc source at test point 5E.

Q.3. Now let us return to the point at which we connected a jumper from the common terminal of the video-output emitters (5M, Figure 14–2) to the chassis. Did brightness return?

A.3.Y. Yes? We can conclude that the video-driver panel (L) is defective or we have a loss of 35 V dc at point 7L (Figure 14–2).

Q.4. Now return to the point at which we bridged a 10,000-ohm resistor from the collector to the base of the green video-output amplifier, test point 6M to 1M (Figure 14–2). Does a green raster appear? (Remember, the 10,000-ohm resistor must be removed before proceeding.)

A.4.Y. Yes? Bridge a 22,000-ohm resistor from the base circuit of the green video drivers to the chassis (14L, Figure 14–2, to ground). This resistor bypasses some of the color-drive signal to eliminate this as a source.

Q.5. Does brightness return? (Remove the 22,000-ohm resistor before proceeding.)

A.5.N. No, brightness does not return. Check the voltage at the emitter of the brightness stabilizer (*Q*7E, Figure 14–2). Is this voltage greater than 0.9 volt? If so, we can conclude that the video amplifier (panel E) is defective. If not, we can conclude that the video-driver panel (L, Figure 14–2) is defective or there is a loss of the 35-V dc source at test point 7L (Figure 14–2).

A.5.Y. If brightness did return, bridge a jumper across the second color-IF secondary test points 12S to 13S (Figure 14–2). Does brightness return? If the answer is yes, we can conclude that we have a defective color panel (S, Figure 14–2). If the answer is no, we conclude that our contrast control or related components are defective.

At this point it is necessary to review some basic picture-tube theory. First, there must be high voltage fed to the second anode (CRT high-voltage connection). Second, there must be accelerating voltages at the screens (G1 and G2). Also, the grids must be positive with respect to the cathodes to signal the guns on. Therefore any defects in the drive circuits that signal the electron guns to off will cause a lack of raster even though the accelerating voltages are available. First determine if the accelerating voltages are present. Then isolate the circuits that are causing the improper bias of the grids and cathodes of the picture tube, creating a lack of brightness.

Inadequate Height

Q.1. Adjust vertical linearity and height controls for best linearity and height. Set the voltmeter selector to the ac scale. Measure the peak-to-peak voltage at the base of the vertical-output transistor (*Q*1H). Is the ac voltage at least 8.5 V peak-to-peak? (See Figure 14–3 for location of this test point.)

A.1.Y. Yes? We can conclude that the convergence panel (H) is defective.

A.1.N. No? Unsolder the collector of the vertical current limiter (*Q*2H) which is located on the convergence panel, (Figure 14–3). Does vertical height return? Yes? We can conclude that the convergence panel (H) is defective. If the vertical height does not return, set the voltmeter selector switch to the positive dc scale. Measure the source voltage at 16E, (Figure 14–2) of the video amplifier (E). This test point is at the base of the vertical driver. Is the voltage at least 85 V dc? If so, we can conclude that the video-amplifier panel (E) is defective. If not, we can conclude that the power supply is defective.

Poor Focus

Q.1. Rotate the focus control through its range slowly while observing screen. Does focus improve between both extremes of the focus control?

A.1.Y. Yes? We can conclude that the CRT is defective.

A.1.N. No, it does not improve at either end of the focus control. Then does the focus control have appreciable effect on focus? If so, check or replace the high-voltage rectifier. If not, short the focus-control slider to the other active terminal on the control. Does this affect focus? Yes? We can conclude that the focus control is defective No? Check the focus plug and CRT socket. If that is not defective, replace the focus resistor block (160-Megohm focus-dropping resistor).

Little or No Vertical Sweep

Q.1. Rotate the vertical-centering control through its range. Is there any movement of the horizontal line in the vertical direction?

A.1.N. No? Check the yoke plug. If it is all right, bridge a jumper from terminals 8G to 10G (Figure 14–2). This is the secondary of the vertical pincushion transformer. Does this restore vertical sweep? If so, the pincushion panel (G) is defective. If the answer is no, the deflection yoke or connections to the yoke are defective.

A.1.Y. If there is movement of the horizontal line in the vertical direction, replace the vertical-output transistor (*Q*1H, Figure 14–3). Rotate the vertical size control through its range rapidly. Does this cause momentary vertical sweep?

If so, the video-amplifier panel (E, Figure 14–2) is defective or there is a loss of source voltage to 16E or 5E, shown in Figure 14–2. The video-amplifier panel (E) contains the vertical-discharge and vertical-driver circuits. (We have already shown that the vertical amplifier was operable.) If there is no momentary vertical sweep, unsolder the collector lead at the vertical current limiter (*Q*2H, Figure 14–3). Does this restore vertical sweep?

If it does, the convergence panel (H) is defective. If it does not restore vertical sweep, connect a jumper from terminal 1H to the chassis (vertical-output emitter to ground). If this restores vertical sweep, convergence panel (H) is defective. If not, measure the voltage at the collector of the vertical-output transistor *Q*1H. Is this voltage more than 30 V dc?

If the voltage is more than 30 V dc, the video-amplifier panel (E) is defective or there is a loss of source voltage at terminals 5E or 16E. If the voltage is not more than 30 V dc, the convergence panel (H) is defective.

As in the tests previously used, eliminate the circuits that could load, or find the circuits that do not provide, the correct drive signals.

No Vertical Sync, Normal Horizontal Sync

Turn the noise-threshold control fully counterclockwise. Set the AGC control to midrange. The noise-threshold control is located in the base circuit of the noise-separator transistor (*Q*3E, Figure 14–3). The AGC control is located in the emitter circuit of the AGC-gate transistor (*Q*5E, Figure 14–3). Both transistors are located on the video-amplifier panel (E). Replacing this panel should restore vertical sync.

No Horizontal Sync, Normal Vertical Sync

Q.1. Adjust the horizontal-hold control. Can the horizontal blanking bars be made to stand-up vertically?

A.1.Y. Yes? Turn the noise-threshold control counterclockwise. Set the AGC control to midrange (both of these controls are located on panel E). If there is no horizontal sync then replace the horizontal panel (F).

A.1.N. No, the horizontal blanking bars cannot be made to stand-up vertically. Replace the horizontal panel (F) for it is defective.

You might question why the replacement of the horizontal panel was not suggested immediately. The noise threshold and AGC control must be tried because these controls could easily be maladjusted, resulting in no horizontal sync.

No Horizontal or Vertical Sync

Turn the noise-threshold control fully counterclockwise. Set AGC control to midrange. If this does not restore sync, replace the video-amplifier panel (E).

Improper Gray Scale

Prior to the analysis procedure we will describe the black-and-white tracking procedure.

1. With the tint control centered, tune to a black-and-white signal and turn all CRT screen-grid controls and video-drive controls fully clockwise.
2. Reduce the brightness until a raster is just visible. Turn the screen control that corresponds to the predominant raster color fully counterclockwise.
3. Readjust brightness until the raster is just visible. Turn the screen control that corresponds to the predominant raster color fully counterclockwise.
4. Adjust screen control, which is still at maximum clockwise setting, to the center of its range. Leave this control at this setting for the remainder of the setup.
5. Adjust the other two screen controls to produce a white raster at low brightness-control settings.
6. With the contrast and brightness set to maximum, adjust the video-drive controls to produce whites in picture highlights.
7. Again, reduce brightness until raster is just visible. Adjust the screen controls to produce a white raster. (Repeat steps 5 through 7.)

This tracking requirement assures a uniform black-and-white picture at all brightness levels.

Q.1. Turn all screen grids and video-drive controls to minimum (CCW). Advance each pair of controls (red screen, red drive, blue screen, blue drive, green screen, green drive) to produce individual fields. Is each field approximately equal in intensity? Does each control have approximately the same range?

A.1.Y. Yes, the intensities are equal and so are the ranges of the controls. Perform the black-and-white tracking procedure. If black-and-white tracking is still poor, check the CRT with a reliable checker.

A.1.N. No, the fields are not equal in intensity and the controls do not have the same range. Advance all screen controls to maximum (CW). Set the tint control to midrange. Measure the dc voltage at the center terminal of each screen control. Is each reading approximately 600 V dc?

If not, the control or related components associated with the incorrect voltage measurement is defective. If we do read 600 V dc, set brightness and contrast control to minimum. Remove the video-output panel. Measure the voltages at 1L, 2L, and 3L (test points, Figure 14–3) on the video-driver panel (L, Figure 14–3). Are the voltages within 1 volt of each other?

If so, the video-output panel (M) is defective. If not, the video-drive panel (L) is defective.

No Sound, Picture Normal

Q.1. Disconnect the sound-IF cable from the audio panel (D). Turn the volume control to maximum. With your finger on the blade of a metal screwdriver, touch the blade to the vacated receptacle. Is noise heard from the speaker?

A.1.Y. Yes? The video-IF panel (B) is defective.

A.1.N. If we cannot hear noise from the speaker, we have a defective audio panel (D). Before replacing, check the speaker plugs and audio-output transformer. Substitute the audio-output transistor.

Distorted Sound

Q.1. Tune the receiver correctly to a snow-free, medium-strength station. Substitute an 8-ohm loudspeaker. Substitute the audio-output transistor if sound is distorted. If sound is still distorted, disconnect the sound-IF cable from audio panel (D). Turn the volume control to maximum. With your finger on the blade of a metal screwdriver, touch the blade to the vacated receptacle. Is hum heard from the speaker?

A.1.Y. Yes? We have a defective video-IF panel (B).

A.1.N. If no hum was heard, adjust the top and bottom slugs of sound-IF and ratio-detector coils (*T*1D and *T*2D, Figure 14–3) for best sound. If sound is still distorted replace the audio panel (D).

No Color

Q.1. Short terminal 6F (Figure 14–3) on the horizontal panel to the chassis. This is the base of the color-gate pulseformer. Does color appear on the screen?

A.1.Y. Yes? We can conclude that the horizontal panel (F) is defective.

A.1.N. If color does not appear on the screen, unsolder the collector lead of the color-gate pulsetransformer (*Q*8E, Figure 14–3) on horizontal-sweep panel (F). Does color appear on the screen? If so, the horizontal panel (F) is defective. If not, the color panel (S) is defective.

No Color Sync

Set the horizontal-hold control for best sync. Then adjust the color-oscillator coil for best color sync. Adjust the color-sync filter transformer for minimum color on the screen. If color sync is still poor, the color panel (S) is defective.

One Primary Color Absent

If one primary color is absent, the video-drive panel (L) is defective. This panel is the video drive for each color signal.

One Primary Color Predominant

If one primary color is predominant, the video-drive panel (L) is defective.

questions

1. Name the preliminary steps to be taken before proceeding to isolate a suspected panel.
2. What panel or panels can cause a loss of audio?
3. Does the black-and-white signal pass through the color panel?
4. Name the panels in the black-and-white signal path that could cause a loss of raster.
5. What panels could cause a loss of vertical sweep?
6. What panel is likely to cause a loss of vertical sync?
7. Name the panels or panel that could cause absence of color when turned to a color telecast.
8. What panels can cause a loss of high voltage?
9. What panels can cause excessive brightness?
10. Improper gray scale describes a deficiency in black-and-white tracking. True _____ False _____
11. Make a list of five picture symptoms and a possible location of the trouble for each.

15 out-of-set panel analysis

Introduction

Out-of-set panel analysis is intended to help technicians service inoperable panels on the bench, away from the main chassis. The advantages are understandable. It is easier to deal with a small panel than with a complete television chassis. The professional technician can exercise an extra option. He can replace a complete panel, or he can repair it.

Some out-of-set procedures have no precedent in vacuum-tube circuits but make servicing transistor circuits gratifying. For example, it is possible to connect a 12-volt battery to a test point on an out-of-set panel and observe that the circuits are operating normally even though the rest of the receiver is miles away in the customer's home. Not only that, many panels have ready-made signal sources which can be used on the same panel, or others.

To make out-of-set panel analysis a success, it is necessary to observe a few ground rules.

1. Proceed on the basis that the panel to be analyzed has been replaced by a good one, restoring the receiver to normal operation.
2. The panel to be analyzed must be identified with symptom. Not knowing the symptom places extra weight on the technician and defeats the out-of-set technique.
3. A knowledge of basic transistor theory and practice is necessary to accomplish the desired objective.

There are limitations to this approach and we should be familiar with them. Whether or not out-of-set panel analysis is effective depends upon how much the technician knows about his work. No book or training program can possibly anticipate every possible fault. There will be times when the serviceman will need to know more than how to simply hook up a voltage source and jumper a few test points. Also, there are some defects that cannot be discovered with the panel on the bench. The defect might be a chassis component or involve an ac component that requires full voltages and signals for evaluation.

A 12-volt battery and a few resistors and jumpers will help locate most defects, but no universal remedy is implied. Accept the method for what it is—a way to service a panel out-of-set that is *effective most of the time.*

The voltages in the foldout schematic reflect actual set operation. The voltage readings we will discuss reflect what happens when the panel is out of the set and energized with a 12-V dc source. When looking at the schematic for any reason, ignore the voltages specified unless the panel is in a set.

The Video-Amplifier Panel (B)

We will read voltages on a properly biased panel. An incorrect voltage must be understood in terms of basic, practical transistor theory and corrected accordingly. When all transistors are operational on dc, not much more can be done without the proper equipment. If you are not equipped with a signal generator, you may build the one shown in Figure 15–1.

The signal generator develops a 45-MHz carrier frequency and is modulated with a 5-kHz square wave. The signal can be inserted at the input to the video-amplifier panel and observed at the output terminal 2B (Figure 14–2). Ahead of the video detector (*E*1B, Figure 15–2), a demodulator probe must be used at the scope. The 45 MHz cannot be seen by a scope, so it must be detected by a probe. Another alternative is to let the video act as the scope probe. Rather than moving the scope from point-to-point, insert the signal from stage-to-stage, leaving the scope on terminal 2B (Figure 14–2). Each technique will work. Start with the third video-IF and move toward the first with the signal lead. A dead stage won't pass the signal for the scope to display.

When constructing the circuit in Figure 15–1, make sure the leads are as short as possible. Use a shielded output lead, and house the generator in a shielded box. Use one ground point for all grounds. Without these precautions, stray coupling and radiated signals would make the exercise worthless.

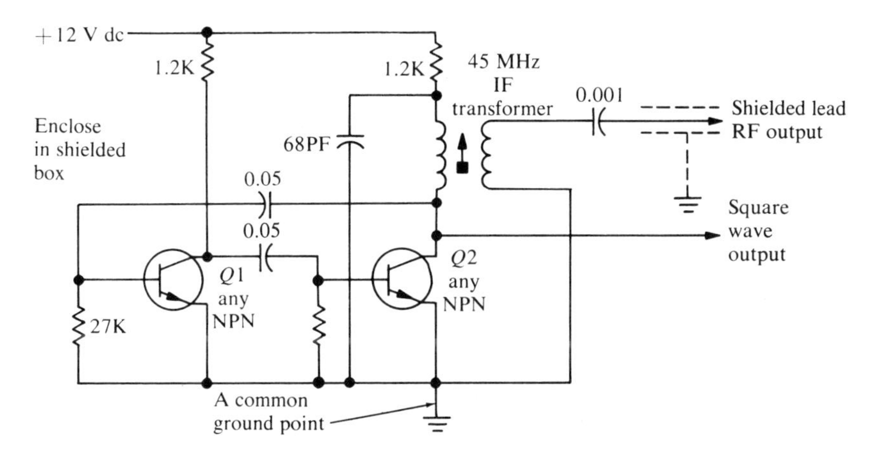

Figure 15–1

Figure 15–2 illustrates the simplified block diagram of the video-IF panel (B). Five closely related functions are performed on the video-IF panel:

1. Picture and sound-IF carrier amplification
2. Video detection
3. Sound heterodyne detection
4. Video amplification
5. AGC amplification

Three transistors *Q*1B, *Q*2B, and *Q*3B provide the amplification required to drive the video detector, *E*1B (Figure 15–2). Rejection of

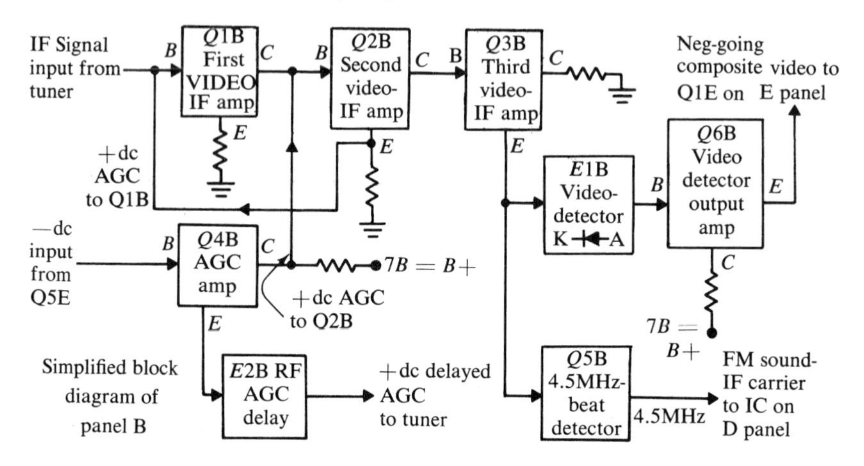

Figure 15–2

frequencies beyond those that constitute the desired picture content is a function of trap circuits, which reject specific undesired frequencies and fix IF amplifier selectivity. Bridge neutralization makes each transistor a one-way device, preventing output signals from becoming part of the input. Voltage dividers set a reduced source voltage for each collector. One must watch these dividers very closely when analyzing individual stages because an open or cutoff transistor will show a collector at the divided source, rather than actual source voltage, unless there is a defect in the divider. Also, the divider resistors are bypassed. Open bypasses permit ac signals across the dividers. This reduces gain and can introduce spurious beats and tweets. Shorted bypass capacitors ground the divider resistors, which lowers or removes the collector voltage and possibly causes the divider to overheat and burnout.

The Video Detector (E1B)

Composite video voltages which make up picture elements are stripped off the picture carrier at the video detector *E*1B (Figure 15–2). The sound-IF carrier is trapped out hard before the detector to prevent a beat between the picture and sound-IF carriers. The video detector is lightly forward-biased so no signal voltage is lost switching on the diode. This bias also makes detection a more linear function, because no voltage is lost in the "knee" of the diode's initial conduction characteristic.

The Heterodyne Detector (Q5B)

A heterodyne (beat) is developed at the heterodyne detector (*Q*5B Figure 15–2) by mixing the sound-IF and picture carriers. A new frequency, (4.5 MHz) results, which is the difference between the two carriers and contains the modulating frequencies of the original sound carrier. The 4.5-MHz sound carrier is coupled to the audio panel (D) via shielded cable from terminal 5B (Figure 14–3).

The Video-Detector-Output Amplifier (Q6B)

The low base/emitter junction impedance of a transistor acts as a load on any circuit to which it is connected. To prevent loading the video detector and losing most of the signal, the video-detector-output amplifier (*Q*6B, Figure 15–2) is arranged in an emitter follower pattern. Composite-video voltage is presented to the base and taken off the emitter. The higher input impedance results in negligible signal voltage loss. The signal is available at terminal 2B for the trip to the video-amplifier panel (E, Figure 14–2).

The AGC Amplifier (Q4B)

The AGC amplifier acts as a resistance which is varied by a voltage. The higher the control voltage, the lower the resistance. The control voltage is a result of conduction of the AGC gate (*Q*5E, Figure 15–2) and is proportional to signal strength. Therefore, the stronger the signal, the higher the control voltage. With a higher control voltage at the input, the AGC amplifier conducts hard. Its low resistance in this state permits an AGC voltage to be applied to the second video-IF amplifier (*Q*2B, Figure 15–2). The result is to hold constant the gain of the IF system with changes in signal strength.

Video-Amplifier Panel (B) Analysis

Now we will show how to isolate an inoperative stage of the video-amplifier panel (B) by checking dc voltages. The instruments required for this task, other than the 12-V dc source are: a 2,200-ohm ½-watt resistor, a 10,000-ohm ½-watt resistor, a 47,000-ohm ½-watt resistor, and a 22,000-ohm ½-watt resistor. The preliminary steps that must be taken are as follows:

1. Hook up the 12-V dc supply to test point 7B.
2. Clip a jumper on 7B and leave one end free.
3. Tack on a 2,200-ohm resistor on test point 2B leaving one end free (this point is located at the emitter resistor of the video-detector-output amplifier).

Q.1. Read the collector voltage at the first video-IF amplifier (*Q*1B, Figure 15–2). Is it about 8.2 V dc?

A.1.N. No, it is not 8.2 V dc. Replace *Q*1B (Figure 15–2) and *R*4B (see foldout schematic). If the collector voltage is missing or low, remove and test *Q*1B. With *Q*1B removed, check the collector voltage. If it is still low, replace *C*11B, *C*12B, *C*13B, *T*1B, *C*15B, and *C*9B (see foldout schematic), one at a time, in that order. Reinsert *Q*1B. Normal voltages should be:

Emitter = 0 V
Base = 0 V
Collector = 8 V

A.1.Y. Yes, it is about 8.2 V dc on the collector. Take the free end of the jumper (on 7B) and touch it to the free end of the 2,200-ohm resistor on test point 2B. Does the collector voltage drop to about 1.5 V dc? If not, replace *Q*1B and *R*4B. If the collector voltage does drop to about 1.5 V dc, we can conclude that the first video-IF amplifier (*Q*1B) is operable as far as the dc paths are concerned.

Q.2. Read the collector voltage at the second video-IF amplifier, (*Q*2B, Figure 15–2). Is it about 8.2 V dc?

A.2.N. No, the collector voltage on the second video-IF amplifier (*Q*2B) is not about 8.2 V dc. Replace *Q*2B (Figure 15–2) and *R*8B (see foldout schematic). If the voltage is low or missing at the collector, remove *Q*2B and test it. While it is removed, check the collector voltage. If it is still low, replace or check *C*22B, *C*19B, *C*21B, *C*16B, and *T*2B (see foldout schematic).

Reinsert the good second video-IF amplifier (*Q*2B). The normal voltages should be:

Emitter = 0 V
Base = 0.6 V
Collector = 8.2 V

Q.3. Touch the jumper to the 2,200-ohm resistor. Does the collector voltage drop to about 2.4 V dc?

A.3.N. No, the collector voltage does not drop to about 2.4 V dc. Replace *Q*2B (Figure 15–2) and *R*7B (see foldout schematic).

A.3.Y. Yes, the collector voltage does drop to about 2.4 V dc. The second video-IF amplifier (*Q*2B) is operable as far as the dc path is concerned.

Q.4. Read the collector voltage at the third video-IF amplifier (*Q*3B, Figure 15–2). Is it about 9.6 V dc?

A.4.N. No, the collector voltage at the third video-IF amplifier (*Q*3B) is not 9.6 V dc. If it is much higher, *Q*3B might be cutoff. Touch a 10,000-ohm resistor from test point 7B to the base of *Q*3B. The collector voltage should drop to 5.6 V dc. If it does not, replace *Q*3B (Figure 15–2) and *R*14B (see foldout schematic). If it does, replace *R*13B and *R*38B (see foldout schematic). If the collector voltage is low or missing, remove the third video-IF amplifier (*Q*3B) and test it. While it is removed, check the collector voltage. If it is low, check or replace *C*28B, *C*27B, *C*25B, and *C*24B (see foldout schematic). Reinsert the good *Q*3B transistor. Check the voltages again. Normal voltages should be:

Emitter = 0.1 V
Base = 0.6 V
Collector = 9.6 V

A.4.Y. Yes, the collector at the transistor of the third video-IF amplifier (*Q*3B) measures about 9.6 V dc. Short the emitter to its base. Does the collector voltage rise from 9.6 to 11 V dc? If so, it can be concluded that the third video-IF amplifier (*Q*3B) is operable as far as the dc paths are concerned. If not, replace it and *R*14B. (See foldout schematic).

Q.5. Read the collector voltage at the 4.5-MHz beat detector (*Q*5B, Figure 15–2). Does it measure about 6 V dc?

A.5.N. No, the collector voltage at the 4.5-MHz beat detector (*Q*5B) does not measure 6 V. Replace *Q*5B (Figure 15–2) and *R*34B (see foldout schematic). If the voltage is low or is missing, remove the 4.5-MHz beat detector (*Q*5B) and test it. Read the collector voltage. If it is still low, check or replace *C*36B, *C*44B, *C*46B, *C*47B, *L*10B, *C*42B, and *R*31B (see foldout schematic). Reinsert the good transistor Q5B. The normal voltages should read as follows:

Emitter = 0 V
Base = 0.6 V
Collector = 0.6 V

A.5.Y. Yes, the collector voltage does measure 6 V at the 4.5-MHz beat detector (*Q*5B). Bridge a 47,000-ohm resistor from the collector to its base. Does the collector voltage drop to 3 V dc? If not, replace *Q*5B and *R*34B and check *R*33B and *R*32B (see foldout schematic). If so, it can definitely be concluded that the 4.5 MHz beat detector (*Q*5B) is operable as a dc amplifier.

Q.6. Read the collector voltage at the video-detector output amplifier (Q6B Figure–2). Does it read about 10.5 V dc?

A.6.N. No, it does not read about 10.5 V dc on the collector of *Q*6B. If the reading is high, bridge a 10,000-ohm resistor from test point 7B to base of transistor *Q*6B. The collector voltage should drop to 6.6 V dc. If it does not, replace the video-detector-output amplifier (*Q*6B), *R*40B, and *C*49B (see foldout schematic). If it does, replace *R*41B and *C*48B (see foldout schematic). If it reads low or the voltage is missing, remove *Q*6B and test it. Check *R*41B, *C*48B, *R*51B, *R*20B, and *R*41B; *C*48B, *R*51B, *R*20B, and *R*18B (see foldout schematic). Reinsert the good transistor *Q*6B. The normal voltages should be:

Emitter = 1.2 V
Base = 1.8 V
Collector = 10.5 V

A.6.Y. Yes, it does read about 10.5 V dc on the collector of the video-detector-output amplifier (*Q*6B). While reading collector voltage at *Q*6B, short the emitter to the base of this transistor. Does the collector voltage rise to 12 V dc? If not, replace *Q*6B and *R*40B. If so, it can be concluded that the video-detector-output amplifier is operable as far as the dc paths are concerned.

Q.7. Read emitter voltage at the AGC amplifier (*Q*4B, Figure 15–2). Is it about 12 V dc?

A.7.N. No, it is not 12 V dc on the emitter of *Q*4B. If it is low or missing, check *L*9B, *C*37B, *C*40B, and *C*23B (see foldout schematic). Remove and test *Q*4B. Check the emitter voltage. If it reads low, check *C*39B (see foldout schematic). Reinsert the good AGC amplifier. If the emitter voltage is still low, short the emitter to base. Emitter voltage should rise to source. If it does not, the defect is in the emitter circuit. If it does, the defect is in the collector circuit. Check *C*41B, *C*50B, and *C*18B (see foldout schematic). The normal voltages should be:

Emitter = 12 V
Base = 12.2 V
Collector = 6 V

A.7.Y. Yes, it does measure about 12 V at the emitter of *Q*4B. While reading the collector voltage, bridge a 22,000-ohm resistor from the collector to the base. Does the collector voltage rise from 0.5 V dc to 6.4 V dc? If not, we can conclude that we must replace *Q*4B (Figure 15–2), *R*27B, and *R*24B (see foldout schematic). If so, we can conclude that the AGC amplifier is operable as far as the dc path is concerned.

Check *E*1B (diode in second detector, Figure 15–2) with an ohmmeter. Forward resistance ($R \times 100$) should be about 350 ohms. Reverse resistance ($R \times 100$) should be 3,500 ohms.

If all devices are cleared as operational on dc, any remaining difficulties will be caused by ac defects. These will be difficult to analyze without a signal source. Tweets, poor resolution, weak or excessive contrast, poor sensitivity, and regeneration are best analyzed with the panel mounted on a receiver chassis.

Assuming that the dc checks are all right, there are some ac general trouble spots that can be suggested:

1. Open neutralizing and emitter bypass capacitors cause weak contrast and poor sensitivities (*C*8B, *C*10B, *C*12B, *C*15B, *C*17B, *C*19B, *C*20B, *C*21B, *C*25B, *C*26B—see foldout schematic).
2. Open coupling capacitors cause complete signal loss (*C*1B, *C*6B, *C*16B, *C*24B—see foldout schematic).
3. Open decoupling capacitors cause beats, tweets, poor sensitivity, weak contrast, and apparent misalignment (*C*11B, *C*13B, *C*22B, *C*23B, *C*28B, *C*40B—see foldout schematic).
4. Open electrolytics cause beats, tweets, weak contrast, poor sensitivity, and apparent misalignment (*C*37B, *C*38B—see foldout schematic).

The Audio Panel (D)—Integrated Circuit (IC)

The audio panel contains a straightforward audio amplifier (*Q*3D, Figure 15–3) and a driver stage (*Q*4D, Figure 15–3) which amplify detected audio signals sufficient to drive the chassis-mounted audio power-output transistor (*Q*5D, Figure 15–3), transformer, and loudspeaker.

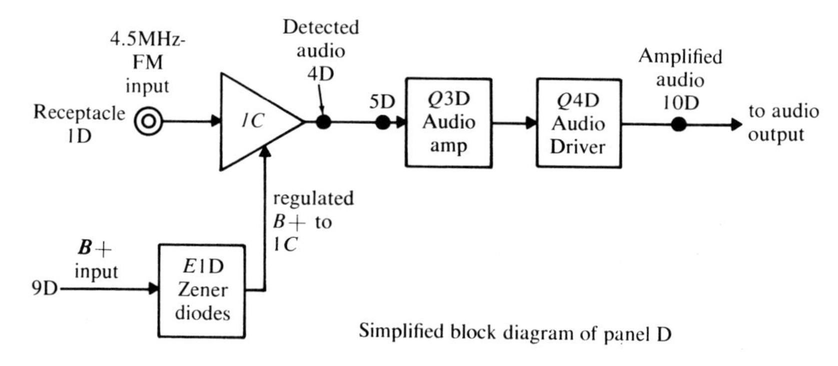

Simplified block diagram of panel D

Figure 15–3

We will not go through the same technique that was performed on Panel (B) for the amplifier and the driver. The same basic technique is used, but we will assume that these stages are operating as required.

Instead we will involve ourselves with the method used to test Zener diodes and the integrated circuit. The integrated circuit (*IC,* Figure 15–3) performs four functions:

1. 4.5-MHz FM carrier amplifier
2. 4.5-MHz FM carrier AM limiting
3. 4.5-MHz carrier audio detection
4. Audio preamplifier

Since all these functions are housed in one device, the following sound symptoms can be related to the *IC:* no audio, noisy audio, weak audio, distorted audio, audio drift, and poor audio sensitivity.

Inside the *IC*; 12 transistors are direct-coupled (no coupling capacitors) to perform the various functions. Also involved are 12 diodes and 16 resistors. A short or open in one of these internal components will affect total *IC* current drain. We will suggest several ways current drain can be checked. Significant deviations from listed values lead to a legitimate suspicion that the *IC* is incapable of performing one or more of its functions and should be replaced. It is hard to see how an *IC* with an internal defect would not also affect current drain. Anything is possible, however, and a dynamic check of the *IC* capabilities is the only real way to determine if the *IC* is doing everything it is supposed to do. Feeding a frequency-modulated 4.5-MHz carrier into the input and measuring the recovered audio at the output is a more positive way out of the mystery.

Finally, transformers *T*1D and *T*2D play important roles in *IC* performance. Here again, the best check is obtained with signal injection.

Servicing the Audio Panel (D) (Zener Diode)

Again we need the use of a 12-V dc supply. To test this panel we need a 120-ohm ½-watt resistor with clips on each side of the resistor. We connect the 12-V supply to test point 9D on panel (D). A jumper is tied across the source-dropping resistor *R*17D (see foldout schematic).

Q.1. While measuring voltage across the Zener diode *E*1D (Figure 15–3), bridge a 120-ohm resistor across *R*4D (see foldout schematic). Does Zener voltage rise above 10 V dc?

A.1.N. No, the Zener voltage does not rise above 10 V dc. We can conclude that if the Zener diode holds the source voltage to 10 V or less it is functioning normally.

A.1.Y. Yes, the Zener voltage does rise above 10 V dc. We can definitely conclude that the Zener diode *E*1D is defective.

Servicing the Integrated Circuit (IC)

Q.1. With 12 V dc supplied to test point 9D, measure the voltage drop across *R*17D (see foldout schematic). Is it approximately 5.5 V dc?

A.1.N. No, it is not approximately 5.5 V dc across *R*17D. An abnormal drop across *R*17D is most likely caused by the integrated circuit (*IC*), Figure 15–3. We know that the audio amplifier (*Q*3D) and audio driver (*Q*4D) shown in Figure 15–3 are functioning normally.

A.1.Y. Yes, the voltage drop is approximately 5.5 V dc. We can conclude that the *IC* is not shorted, open, or leaky.

Q.2. Measure voltages at the *IC*. Pins 1, 2, 3, 4, 6, 7 should read approximately 1.5 V dc. Pins 5 and 10 should read approximately 4 V dc. Pin 8 should read 8 V dc and pin 9 should read 0.3 V dc.

A.2.N. No, the voltages do not vary from those specified. We can conclude that the *IC* is not shorted, leaky, or open.

A.2.Y. Yes, the voltages vary. We can suspect the *IC*. To prove whether it is faulty we must unsolder lead 8. Connect a millimeter from the unsoldered lead to ground. Normal *IC* draws approximately 7.5 mA. Replace the *IC* if the current is high or lower than specified.

We can check the *IC* dynamically by connecting a VTVM to test point 4D (Figure 14–3). Connect an FM signal generator to the input jack and set it at 4.5 MHz. Modulate the RF carrier with a 400-Hz signal and set the carrier deviation to 7.5 kHz. Set the output of the generator to a level of 100 mV rms. Read the recovered audio on an ac VTVM at test point 4D. It should approximate an output of 45 mV rms.

The Horizontal-Sweep Panel (F)

Seven transistors and an AFC (automatic frequency control) device make up the horizontal-sweep function. The F panel (Figure 15–4) is an ac generator which produces a rectangular voltage signal at a 15,750-Hz basic frequency. The deflection yoke inductance creates a sawtooth current from this signal. The yoke field moves the CRT beams horizontally at a line-scan rate.

The rectangular signal is also applied across the primary of a high-voltage transformer (not on the panel) which steps up the peak-to-peak voltage across the secondary. Here the ac signal is rectified by a vacuum tube. The dc voltage (about 25-27 kV) is connected to the second anode of the CRT. A pulse coil wound on the high-voltage transformer core applies pulses to a two-diode centering network mounted on the F panel. A pulseformer circuit turns out a gating pulse used by the gated color-sync amplifier (*Q*7S, see foldout schematic) on the color panel (S, Figure 14–3).

The Horizontal Oscillator (Q2F)

The horizontal oscillator is an off-on switch connected from source to ground. When the switch is closed (on), the collector voltage is shorted to ground. When the switch opens (off), the collector voltage rises to source value. This up-and-down collector voltage is a rectangular signal when looked at with an oscilloscope. The switch (*Q*2F, Figure 15–4) is turned on and off by the oscillator coil (*L*1F, see foldout schematic) and related components.

The Predriver (Q3F)

The predriver (*Q*3F, Figure 15–4) is switched on and off by the preceding collector at the horizontal oscillator (*Q*2F). When the oscillator is on, the predriver is also on. The prediver is off when the oscillator is off.

The Driver (Q4F)

The driver (*Q*4F, Figure 15–4) is also a switch which is open and closed at a 15,750-Hz line rate. When the horizontal oscillator and predriver are on, the driver is switched on, shorting the collector voltage to ground. When the horizonal oscillator and predriver snap open, the driver shuts off, too. The unloading of the driver transformer (*T*1F, Figure 15–4) induces a pulse which is dampened by an rc network from collector to ground. The positive-going pulse is coupled to the secondary and applied to the output transistors as an on signal.

The Output Pair (Q6F and Q7F)

Operating in parallel, the output pair (*Q*6F and *Q*7F, Figure 15–4) are switches across a regulated collector source (74 V dc in the set). When the switches are closed (on), the supply is grounded. When open (off), the supply rises 9 or 10 times beyond its initial value (750 to 800 V p-p). This is done by the inductive yoke and high-voltage transformer. This rise-and-fall signal is used by the yoke and high-voltage transformer as an ac source. The induced pulse, created by unloading the inductive components when the output pair are open, is held to about 480 V p-p by a pulse-limiter diode (*E*1F) and an rc network. A damper diode (*E*2F) loads the yoke during the first part of horizontal scan to prevent ringing. (See Figure 15–4.)

The Arc Gate (Q5F)

The arc gate (*Q*5F, Figure 15–4) is a switch which is turned on if an arc occurs in the high-voltage rectifier. In this event, the arc gate switches the horizontal driver on, also. In this saturated condition, no signal is applied to the horizontal output pair. Normally the arc gate is cutoff.

The Color-gate Pulseformer (Q8F)

The color-gate pulseformer is a normally-closed (saturated) switch. Received horizontal sync kicks the switch open. When the collector voltage rises to its source (which is a feedback pulse of relatively long time compared to sync pulse passes), the color-gate pulseformer returns to its normally-closed state. The rise-and-fall of collector voltage makes up a gating pulse required by *Q*7S (see foldout schematic) on the color panel.

Horizontal AFC, (Q1F)

Horizontal AFC (*Q*1F, Figure 15–4) is a symmetrical transistor which is equivalent to two diodes arranged as a phase detector. Two signals are phase-compared at *Q*1F: one, horizontal sync; two, a feedback reference. An in-phase or out-of-phase situation results in a correction voltage for *Q*2F. When the feedback pulse goes through zero during sync time, no correction is required.

To service the F panel, connect a 12-V dc source to test point 7F and 10F (Figure 14–3). Completing the collector circuit for the output pair to test point through a 470-ohm resistor allows all the switches to operate (*Q*2F, *Q*3F, *Q*4F, *Q*6F, *Q*7F). Signal-tracing with a scope, starting at the output pair and going towards *Q*2F, will point out a stage which is not switching (no signal). *Q*8F is checked on a dc basis. *Q*5F is tested for its ability to stop the switching action of *Q*4F. The horizontal-centering diodes, *E*1F and *E*2F, Figure 15–4, are checked with an ohmmeter as is *Q*1F, the horizontal-AFC device. These tests should locate most defects.

Servicing the Horizontal-Sweep Panel (F)

Since the signal source is the horizontal-sweep oscillator on the panel, let's make use of it. In addition to the signal source, a 12-V supply, a 470-ohm ½-watt resistor, and a 47,000 ½-watt resistor should be available.

The 12-V supply is tied to test points 10F and 7F. Connect the 470-ohm resistor from the heat sink to test point 7F. The 470-ohm resistor will provide the collector voltage. Normally this is accomplished by the primary of the flyback (high-voltage transformer). (See Figure 15–4 for the following.)

Q.1. Place the scope input on the heat sink (aluminum mounting for the horizontal-output transistors *Q*6F and *Q*7F). Is the signal present (about 9 V p-p on the VTVM)?

A.1.Y. Yes, the signal is present at the heat sink. We can conclude that the sweep panel is operable. We must then check the AFC, arc gate, and pulseformer.

A.2.N. No, there is not any signal present at the heat sink. Place the scope input on the positive side of *C*13F. (*C*13F is tied to the base of the horizontal output, *Q*7F). Is there a signal present (2 V p-p)? If so, we can conclude that the horizontal-output stage is inoperative. The two horizontal outputs *Q*6F and *Q*7F should be checked. Check diodes *E*1F and *E*2F on the *R* ×

100 scale of the ohmmeter. The forward resistance should read almost open (infinite).

If there is no signal present at the input of the horizontal output, place the scope input on the collector or *Q*4F. Is the signal present (27 V p-p)? If so, the driver transformer (*T*1F) is nonoperable.

Place the scope on the base of *Q*4F. Is there a signal of 0.5 V p-p present? If so, the *Q*4F stage is inoperative. If there is no signal of 0.5 V p-p present at the base of *Q*4F, place a scope on the collector of *Q*3F. Is a signal present of at least 12 V p-p? Yes? Either or both *R*12F or *R*13F resistors are open.

If there is not any signal present on the collector of *Q*3F, place a scope on the base of *Q*3F. Is there a signal present (2 V p-p)? Yes? *Q*3F stage must be inoperable.

If there is not any signal on the base of *Q*3F, place the scope on the collector of transistor *Q*2F. Is a signal of 12 V peak-to-peak present? If not, *Q*2F is inoperable. (We will go into the repair of this stage shortly.)

If there is a signal present on the collector of *Q*2F, we can conclude that *L*2F is open. (Check *R*25F.)

Note: At this point, if we have established that panel F is operable, it is time to proceed to the AFC stage (*Q*1F).

Q.2. Place the scope input on point *B* of *Q*1F. Is there a signal of 0.3 V p-p present?

A.2.N. No, there is not a signal at point *B* of *Q*1F. We can assume that *Q*2F is open or *Q*1F is shorted. Also check *C*3F, *C*4F, and *C*5F. (At this point we have tested *Q*1F and assume that *Q*1F is operable.)

Q.3. Place the scope input on the point *C* of *Q*1F. Is there a signal of 0.3 V p-p present at this point?

A.3.N. No, there is not a voltage of 0.3 V p-p present at point *C* of *Q*1F. We must conclude that *Q*1F is shorted.

A.3.Y. Yes, there is a voltage of 0.3 V p-p present at point *C* of *Q*1F. We can safely assume that *Q*1F is operable. To be certain we can test *Q*1F with an ohmmeter. *Q*8F can be checked at this point since we have satisfied ouselves that the panel is operable.

Q.4. Place the scope on terminal 4F. Is there a signal present of 0.3 V p-p? (This point is on the collector of the pulse-gate pulse-former of *Q*8F.)

A.4.N. No, the signal is not present at test point 4F. We can conclude that *R*26F is open or we have a shorted transistor (*Q*8F).

A.4.Y. Yes, we have a 0.3 V p-p voltage at test point 4F. *Q*8F is operable. Check this transistor with an ohmmeter.

Q.5. Place a scope on the heat sink of *Q*5F. Bridge a 47,000-ohm resistor from the base of *Q*5F to ground. Does the signal collapse?

A.5.N. No, the signal does not collapse. We can conclude that *Q*5F is inoperable. Check *R*15F, *R*18F, *R*17F, *R*16F, and *C*11F.

A.5.Y. Yes, the signal did collapse. *Q*5F is operable.

We tested to see if the oscillator *Q*2F were operable earlier. If we determined that it is not, we must proceed to analyze this stage because of the loss of the signal source.

Horizontal Oscillator (Q-2F)

Connect the 12-V dc supply to test point 10F. (We have definitely proven in the previous tests that the oscillator is inoperable.)

Q.1. Read the collector voltage of *Q*2F. Does it read 12 V dc?

A.1.N. No, the collector voltage of *Q*2F is not 12 V dc. If it is low, *Q*2F is leaky or biased-on. Check *R*8F and *R*7F for correct values with an ohmmeter. If the collector voltage is missing, check *R*10F to see if it is open.

A.1.Y. Yes, there is 12 V dc on the collector of transistor *Q*2F. Bridge the collector to base with a 1,000-ohm resistor. Read the collector voltage at *Q*2F. Is it about 6.5 V dc? If not, we can conclude that *Q*2F is inoperable. Check transistor *Q*2F, *R*9F and *C*5F. If the collector voltage does read about 6.5 V dc, Q2F is operable.

If the *Q*2F device is operable but the stage does not operate satisfactorily, *C*7F and *C*6F must be replaced. Check coil *L*1F for continuity and open or shorted turns. Replace *L*1F if the defect is not found.

The IF panel was selected to illustrate this procedure because of its many ac signal components. The audio panel was selected because of the Zener diode and the integrated circuit it contains. The tests on these panels are unique. The horizontal panel was chosen to demonstrate the use of the circuit signal as a source for troubleshooting assistance.

What about other receivers and other manufacturers' panels? Do we always know what voltages to expect out of the natural receiver environment? This information, if not given, could be obtained by the serviceman. It is a good idea to measure voltages on known good panels out of the receiver and record them.

Remember, this is a suggested method and by no means do we imply that there is no other way. Use your own ingenuity and stay with the method that serves you best. The desired result is to make the defective receiver operable. Whatever method you use, the importance of using the service manual for the specific receiver should never be minimized. The basics do not change, but the methods of obtaining results can differ. With technology in color and black-and-white television progressing as fast as it is, the learning process must also continually progress. The service manual updates our knowledge, as it pertains to a given receiver.

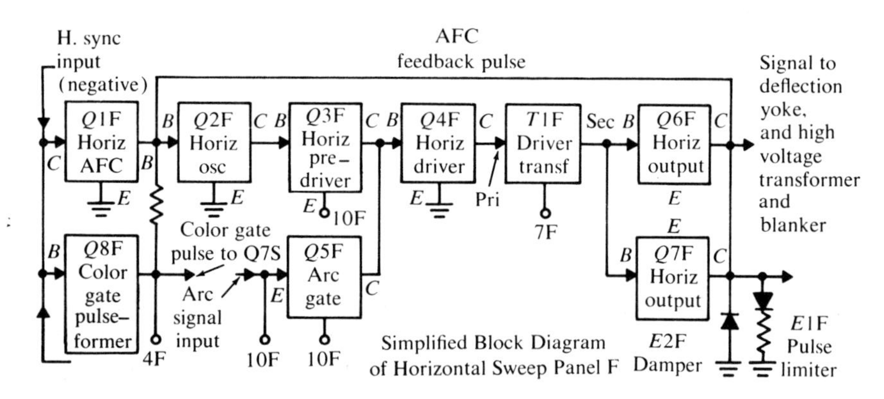

Simplified Block Diagram of Horizontal Sweep Panel F

Figure 15–4

questions

1. What are the five closely related functions performed on the video-IF panel?
2. What two carriers mix to develop the heterodyne (beat) frequency (4.5 MHz)?
3. What is the purpose of the AGC amplifier?
4. Open electrolytics in the IF can cause beats, tweets, ________, ________, and apparent misalignment.
5. The integrated circuit *IC* performs what four functions?
6. How many steps should it take to decide if the Zener diode *E*1D is defective?
7. Can we check the *IC* dynamically with a VTVM? Describe how this could be done.
8. How many transistors make up the horizontal-sweep function?
9. What type of signal is applied across the primary of the high-voltage transformer?
10. What is the function of the arc-gate transistor *Q*5F?

glossary

Alternating Current Reverse direction at regular intervals.

A.M. Amplitude modulation.

Amplitude The height of the alternating voltage or current reached from zero to its most positive point, or from zero to its most negative point.

Amplitude Modulation The envelope of the modulated carrier is the information to be transmitted or received.

Analogy A method of describing a similar action between two different things, using one that is more easily understood.

Anode The element that the electrons are attracted to because of its positive potential.

Aperture Mask Part of a color tube. Also called shadow mask.

Armature The movable portion of a magnetic circuit, such as a generator.

Aspect Ratio A ratio equal to picture width divided by picture height.

Automatic Frequency Control A circuit designed to stabilize the frequency of an oscillator.

Automatic Phase Control In color television receivers, a circuit used to control the frequency, as well as the phase, of the 3.5-MHz oscillator synchronized with the burst signal.

Back Porch The portion on the blanking pulse that contains the 3.58-MHz burst signal of approximately 8 cycles.

Band Frequencies that are within two limits.

Bandpass Amplifier An amplifier designed to amplify the band limits. Term used to define the chrominance amplifier in a color television receiver.

Bar Generator A type of signal generator that delivers frequencies corresponding to several color or chrominance signals of various hues or color phases. The colors appear as bars or bands on the picture tube of color receivers.

Bar Magnet A straight magnet with a north pole and a south pole on opposite ends.

Black Level The base upon which the synchronizing pulses (vertical, horizontal, and 3.58 burst) rest. This is the voltage level that cuts off the flow of electrons, causing the picture-tube face to blacken.

Blue Beam Magnet A circular, small, permanent magnet used to move the electron beam that excites the blue phosphors. It is the static adjustment for the blue beam.

Blue Gun The gun of a three-gun color tube that emits an electron beam for the purpose of exciting the blue phosphors.

Brightness Signal The luminance signal in color television.

Broad Band Amplifier An amplifier designed to provide adequate gain throughout a range of frequencies.

Burst The burst signal originates at the transmitter, and is used in the receiver for synchronizing phase and frequency of the color oscillator.

Burst Oscillator The color oscillator.

***B–Y* Signal** Difference signal which is a component of the chrominance signal. When added to the brightness or *Y* signal, results in the blue primary color signal.

Capacitance The ability to receive or carry a charge of electricity.

Carrier (Frequency) The frequency assigned to the channel (station). It is the frequency of the unmodulated wave.

Carrier Color Signal The chrominance signal.

Cathode-Ray Oscilloscope An instrument that reproduces the wave, shape, or form of the voltage or current.

Channel The range or band of frequencies that are to be transmitted.

Chroma Refers to hue and saturation, does not refer to brightness or luminance. (Has been used to refer only to saturation.)

Chroma Control A color television control that changes the saturation of color.

Chromaticity The quality of a color which depends only on hue and saturation.

Chromatron A single-gun color television picture tube. The Lawrence color tube is such a tube.

Chrominance (See **Chromaticity**)

Chrominance Amplifier The amplifier section of a color receiver that amplifies only the color and burst frequency.

Chrominance Signals Components of a chrominance signal consisting of $R-Y$ and $B-Y$.

Closed Circuit A close loop through which a current flows when voltage is applied.

Coil Turns of wire on a form that may or may not have an iron core inserted into the form. In some cases, the form may be omitted. The coil offers considerable opposition to alternating (ac) current, but not to direct (dc) current.

Color Carrier The composite color-television signal.

Color Carrier Reference The phase of the color-burst signal. This phase synchronizes the transmitter and the receiver.

Color-Difference Signal Color information that is to be transmitted and received. It is composed of $B-Y$, $R-Y$, and $G-Y$ difference signals.

Color Fit Correct registration of the color image with the black-and-white image on the picture-tube screen. This is accomplished by the use of a delay line. The Y signal is slowed up approximately one microsecond by the delay line.

Color Killer Circuit that cuts off the color amplifier during the time monochrome signals are being received. This prevents color interference in the black-and-white picture.

Color Oscillator Oscillator frequency at 3.58 MHz. Its purpose is to furnish a continual subcarrier frequency required for demodulators in the receivers.

Color Phase Color phase determines the hue to which a given signal is related.

Color-sync Signal The burst signal.

Commutator A ring of insulated copper segments connected to the windings of an armature. The brushes of the commutator connect the armature windings to the outside circuits.

Compatability Permits use of a black-and-white receiver to reproduce color pictures on a shades scale from black to white, and permits the color receiver to reproduce black-and-white transmitted signals.

Composite Color Signal All of the color signals transmitted by amplitude modulation. Includes the luminance signal, chrominance signal, sidebands, burst signal, horizontal and vertical and equalizing pulses, together with the blanking pulse.

Condenser Consists of two conducting surfaces separated from each other by an insulating material such as air, oil, paper, glass, or mica. It has the capability of storing electrical energy. Its size is specified in farads, microfarads, or micromicrofarads.

Convergence The superimposing of the three color rasters in a three-gun color tube.

Convergence Coils Three electromagnets carrying currents which produce fields for the convergence of the three color rasters (red, blue, green).

Crosshatch Generator An instrument that produces horizontal and vertical scanning lines. The crosshatch pattern forms precise squares. The raster has a 4 : 3 standard aspect ratio. Most generators of this type also provide a white dot pattern in place of the lines.

Crosshatch Pattern A number of horizontal and vertical lines produced on the screen of a picture tube. This aids in the testing of linearity, size, and convergence. The bars can be white-on-black field or black-on-white field.

Cycle One complete series of changes from zero to a positive peak, back to zero to a negative peak, and returning to zero. The cycle can start at any part, but must complete its 360° in the time period of one cycle.

Definition In television pictures, the degree to which fine details are reproduced.

Deflection Bending of the electron beam, both in the horizontal and vertical direction. The horizontal deflection performs the function of tracing the picture lines while the vertical deflection allows formation of fields and frames.

Deflection of Magnetic Needle Movement of a compass needle caused by magnetic fields or current.

Demodulation Removal of the carrier frequency, so that what remains is the envelope containing the audio and picture information.

Demodulator Circuit that produces color difference signals from variations of phase and amplitude, with reference to an applied color oscillator signal. Chrominance phase varies polarity, while the chrominance amplitude varies the amplitude.

Diode Allows electrons to pass in only one direction, from the cathode to the anode.

Dynamic Convergence Convergence of the deflected beams by the application of magnetic fields created by means other than fixed magnets.

Electromagnetic Flux The flux or magnetic lines of force produced by an electromagnet.

Electron Gun Produces the electron beam and directs it against the fluorescent screen.

Equalizing Pulses Allow alternate picture fields to begin with a full line and a half line for interlace scanning.

Faceplate The transparent glass at the front of the tube through which is seen the phosphor illumination.

Field Every alternate horizontal line of television picture as scanned during 1/60 second. One field begins with a full line from the upper left-hand corner, the next with a half line beginning midway across the top. The two fields take a time of 1/30 second, and form one frame. There are 525 lines per frame, and 262½ lines per field.

Field The space in which appear the magnetic lines of force around a magnet or electromagnet.

Field Flux The lines of force passing through a magnetic field or magnetic circuit.

Field Frequency In interlaced scanning, this term refers to the number of times per second the frame area is fractionally scanned.

Flux The flux in magnetism is similar to current in electric circuits, since both terms refer to flow.

Flux Density The number of magnetic lines per square inch.

Frequency The number of complete cycles per second.

Frequency Response A graph or curve depicting the relative gains of an amplifier at all frequencies.

Green Gun In three-gun color tubes. The electrons emitted from this gun strike only the green phosphor dots of each triad.

***G–Y* Signal** A green minus $-Y$ signal is the color-difference signal for green. When combined with the proper proportion of Y, results in the green color.

Horizontal Sync Correct timing of the receiver horizontal oscillator is achieved with the master oscillator at the transmitter.

Hue The visual color sensation resulting from the particular wavelengths of light, without reference to brightness (100% saturation).

Hue Control Varies the phase of chrominance signals with respect to the burst signal. Usually adjusted for correct flesh tones.

Inductance The characteristic of a coil to oppose any change in current flow. It has no effect on direct current. Inductance is measured in henrys.

Inductive Coupling Magnetic lines of force produced by the flow of current in one coil couple to another coil and produce a flow of current.

In Phase Describes two alternating currents or voltages whose zero and peak values occur at the same time.

Interlaced Scanning Every-other line of the image is scanned during one downward travel of the scanning beam, and the remaining lines are scanned during the next downward travel of the scanning beam.

Interleaving The transmission of the chrominance and luminance signals within the same band of video frequencies. Luminance signals appear around harmonics of the horizontal line frequency, and the chrominance signals appear around harmonics of the color subcarrier frequency.

Intermediate Frequencies The frequency to which all incoming carrier signals are converted before being fed into the IF amplifier.

Lawrence Color Tube A single-gun color television tube (chromatron tube).

Left-hand Rule A method of showing the relative direction of the flux field, current flow in a conductor, and the motion of the conductor through that field. The forefinger points to the direction of flux, the middle finger to the direction of current, and the thumb points in the direction that the conductor is moved.

Line of Force An imaginary line which indicates the direction in which magnetism flows between magnet poles. It is a unit in which magnetic flux is measured.

Luminance Same general meaning as brightness when referenced to color television.

Luminance Channel The channel intended primarily for carrying luminance information.

Luminance Signal Controls brightness, but not color. Has a frequency range of 4.0 MHz, for reproduction of fine details in the pictures. By itself, this signal can produce a monochrome picture.

Luminosity The brightness of a color when compared with white.

Marker Generator A signal generator furnishing oscillator voltage for producing beat frequency marker pips, or containing an absorption circuit for producing breaks or dips on the frequency-response curve of an amplifier or receiver displayed on an oscilloscope. Frequencies on the curve are identified from tuned frequencies of the generator.

Matrix A group of resistors through which the luminance signal combined with signals from the demodulators form the color primary signals and the *G–Y* color-difference signal at the transmitter.

Matrixing The act of combining or mixing chrominance signals with the luminance signals to retrieve the color primary or color difference signals.

Modulation The process of varying the amplitude of the RF carrier signal, or frequency, in accordance with the wave form of the intelligence signal being transmitted.

Monochrome Description of a picture that appears in black, and white, and shades of gray, but has no coloring.

NTSC Nation Television System Committee.

Oscilloscope A test instrument which shows visually on a screen the wave form of a varying current or voltage. The vertical amplifier response must be at least 4 MHz.

Parabolic Current or Voltage Current or voltage whose strength is varied in time with vertical and horizontal deflection. Its shape is the opposite of the convergence errors; thereby, canceling out the errors.

Parallel Resonant Circuit A circuit that is tuned by the use of a coil in parallel with a capacitor. At resonance, it offers a high impedance path to the current flow. This permits a large value of signal voltage to appear across this circuit.

Permeability A measure of the ease with which magnetic flux or lines of force may be established in a magnetic circuit.

Phase Angle The time differences between alternating currents or voltages. The unit of measurement is the degree lead or lag.

Phase Control The hue control on a color television receiver.

Phase Inverter A circuit that changes the phase of the voltage or current by 180°.

Phosphor A fluorescent material used for the screen in a cathode-ray tube. This material becomes luminous when struck by the beam of electrons.

Phosphor Dot The triad of dots in a color tube.

Picture Element The smallest portion of a picture or scene, which is converted into an electrical signal.

Picture Frequency The number of complete pictures which are scanned in one second.

Pincushion Effect Corners are extended outwards, and the sides, top, and bottom curve inward.

Potential A voltage that, compared to a reference point, is more positive or more negative.

Primary Colors Three portions of the wavelengths from the spectrum which are used as primaries and are combined to form all colors or hues for color television. The primaries are blue, red, and green.

Purity The act of the electron beams in a three-gun tube striking only a phosphor for one primary color at one time. Impurity arises when the beam strikes color dots of other primary colors at the same time.

Purity Coil An electromagnet whose field alters the direction of the electron beams so that each beam strikes only the phosphor for one color at a time.

Quadrature A phase difference of one quarter of a cycle.

Raster The phosphor emittence of light when the electron beam strikes the surface. The beam is deflected both horizontally and vertically, and no picture information is present.

Reactance Opposition offered to the flow of alternating current by the inductance or capacity of a part.

Refractive Index The index of refraction of a material is the ratio of the speed of light in vacuum to the speed of light in the material.

Resonance Occurs when the inductive reactance is equal to the capacitive reactance. This causes a balancing or neutralizing of each other, leaving only the resistive component to oppose the current flow.

Resonant Frequency The frequency which produces resonance in a coil-condenser tuning circuit.

Retrace Blanking Voltage pulses for blanking are derived from the vertical-sweep oscillator or the deflection circuit. These pulses are used to darken the picture tube during retrace intervals.

Retrace Lines Narrow, bright, sloping lines. These lines result from the upward travel of the electron beam during vertical retrace periods.

Retrace Period The time during which the blanked scanning beam in a television tube returns to its starting point on the field.

***R–Y* Signal** A red-minus-*Y* or red-minus-luminance (brightness) signal. When combined with the plus luminance $(+Y)$ signal, results in the red color primary signal.

Saturation Describes the amount of white mixed with a pure color.

Scanning The process by which each picture element is reproduced on the screen of the picture tube.

Shadow Mask A thin sheet containing as many small openings as there are groups of (triads) phosphor dots. It is positioned directly behind the screen.

Sinusoidal Having the form of a sine wave.

Sound Carrier A carrier whose frequency is modulated by the audio signal and is located 4.5 MHz higher than the video carrier frequency of the same carrier.

Stray Flux Magnetic lines of force that pass out of the useful magnetic field.

Subcarrier In color television, a signal with a frequency of 3.58 MHz. The chrominance signals modulate this frequency.

Sweep Generators An instrument containing oscillators and associated circuits which furnish voltage varying continually and repeatedly in frequency back and forth through a range called the sweep width. This voltage is used as input for any circuits or devices whose frequency response is to be observed on an oscilloscope.

Sync Pulse In a composite television signal, this pulse rides on top of the blanking pulse. It is used to synchronize the vertical-receiver and horizontal-receiver oscillators.

Trace A visible luminous line produced on the phosphor or viewing screen of a television picture tube by travel of the electron beam over the screen.

Triad The triangle of color dots on the face of the color tube.

Vertical Sync Synchronization of vertical deflection with respect to vertical-sync pulses in the composite signal.

Very High Frequencies (vhf) Frequencies from 30 to 300 MHz.

Vestigal Side Band Television carrier whose side bands extend to a frequency 4.0 MHz above the video carrier and 0.75 MHz below. The 0.75 MHz is the vestigal side band.

Video Applied to signals associated with the picture being transmitted.

Video Carrier The carrier wave whose amplitude is modulated by the video signal, picture variations, blanking and all sync pulses.

Video Signal The changing voltage, which corresponds to the changing lights and shades in the image being scanned at the television transmitter and reproduced at the television receiver.

Wavelength The distance traveled in a time of one cycle by an alternating current, sound wave, or radio wave. For wave motion in air, the wavelength in meters is equal to 299,820,000 divided by the frequency in hertz.

White Level The potential or voltage that causes the brightest or whitest areas in pictures.

Yoke The electromagnetic coils in which sawtooth currents cause deflection of the electron beam. There are a pair of coils for the vertical deflection, and a pair for the horizontal deflection.

Y-Signal The luminance or brightness signal in color television. This signal contains the high-definition details.

appendix A

harmonics

In the section on the interleaving process, the term *energy bunching* was discussed. Energy bunching depends upon the modulation signal being repetitious. A good example of such a repetitious signal is the square wave. A square wave is made up of a fundamental frequency and harmonics of that fundamental. By definition, harmonics are a multiple of any particular frequency. Thus, the second harmonic of a fundamental frequency would be equal to two times that of the fundamental frequency. The complete television signal is a periodic wave whose fundamental frequency is 15,750 Hz. Therefore, the video signal is made up of a fundamental frequency and harmonics of that fundamental frequency. The following pages of this appendix will help to describe the buildup of harmonics.

Figure A–1 illustrates a single cycle of a given fundamental frequency (point *A* to point *C*). Point *A* to point *B* represents one-half of the cycle. The points *A* to *B* will be used to describe the harmonic build up, because point *B* to point *C* repeats in a negative phase.

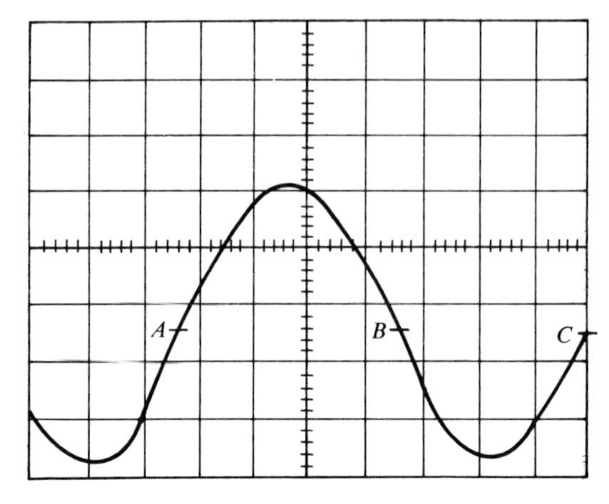

Figure A–1

Figures A–2 and A–3 illustrate the addition of the fundamentals plus the second harmonic. Note: Point *A* to *B* shows that we have a full cycle added to the fundamental. Twice the fundamental (one-half cycle) is equal to a full cycle. It should also be noted that at point *A* we have the positive-going portion, and at point *B*, the negative-going portion of the sine wave. Adding these two frequencies results in a tilted wave shape. The positive-going wave adds to the left side, and the negative-going wave subtracts from the right side, resulting in a tilted wave, as seen in Figure A–3.

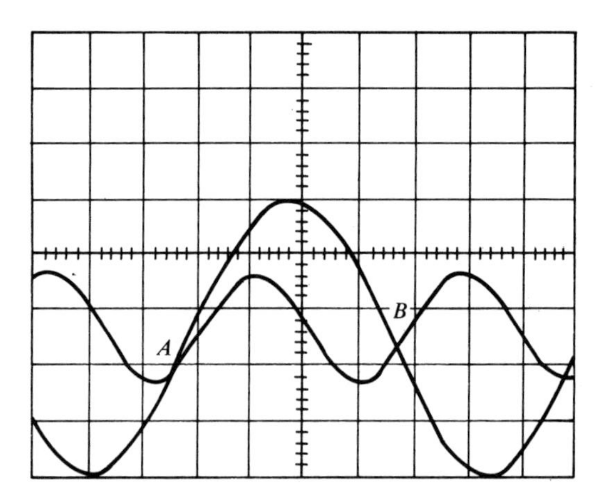

Figure A–2

Figures A–4 and A–5 illustrate the addition of the fundamental plus the third harmonic (one-half cycle times 3 equals 1½ cycles). Note that point *A*

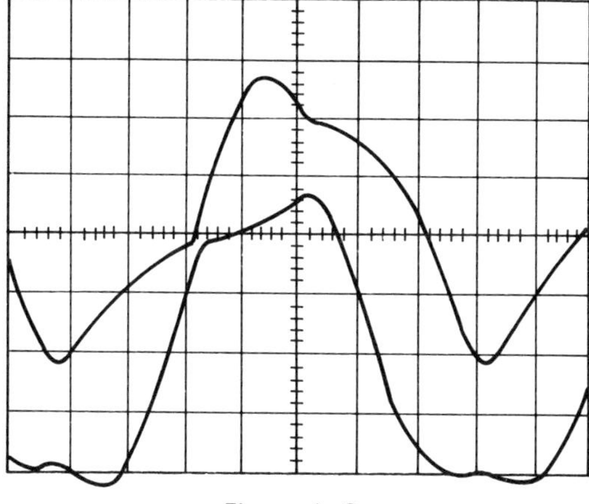

Figure A–3

and point B indicate positive-going waves. These add to the fundamental. The middle portion of the third harmonic wave is negative going and subtracts. This tends to provide equal peaks with a valley in between (Figure A–5).

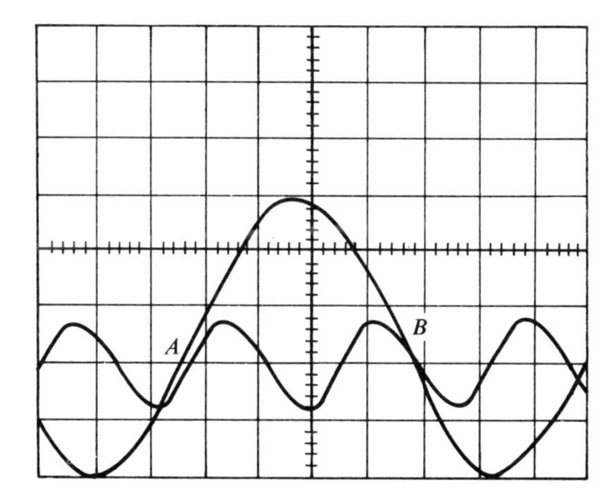

Figure A–4

Figures A–6 and A–7 illustrate the addition of the fundamental frequency plus the fourth (one-half cycle times four equals two cycles) harmonic. Note that at point A the harmonic is positive going, and at point B it is negative going. The center of the fundamental has both a positive and negative signal. Consequently, starting from A to B, we have an increase in amplitude, valley, peak, and dip at the end of this half-cycle time period (Figure A–7). The resultant wave shape is broader and more severely tilted.

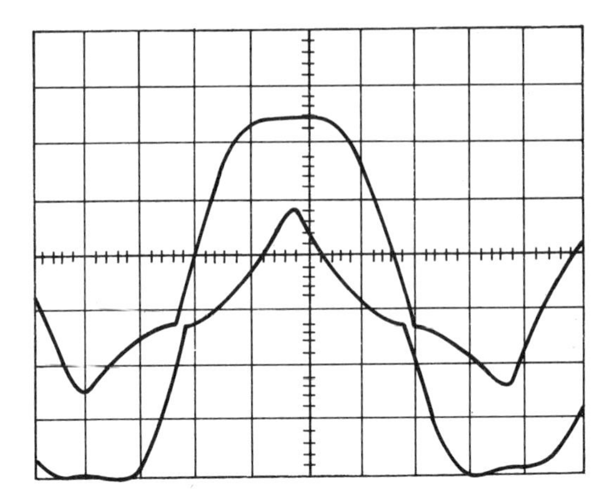

Figure A–5

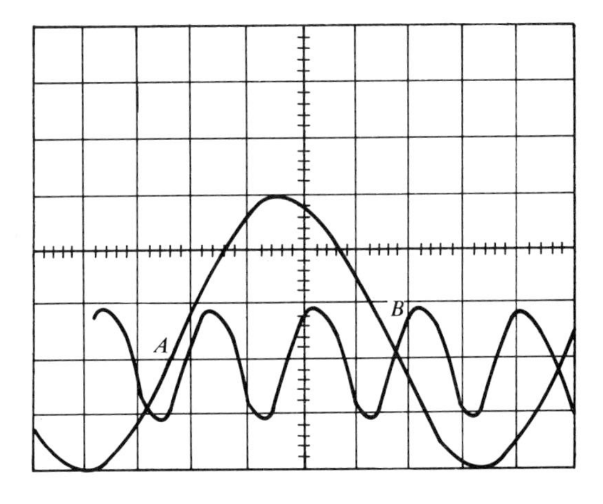

Figure A–6

Figures A–8 and A–9 illustrate the addition of the fundamental frequency plus the fifth (one-half cycle times five is equal to 2½ cycles) harmonic. Note that at point's *A* and *B* the harmonic is positive going. These add to the fundamental frequency at these points. If we traverse from *A* to *B,* we see a peak, valley; peak, valley; and peak.

The resultant wave in Figure A–9 shows that we end up with equal peaks on either end, with a larger center peak. Comparing Figure A–9 to Figure A–5, we see that the result is a broader wave shape with a center peak instead of a center valley.

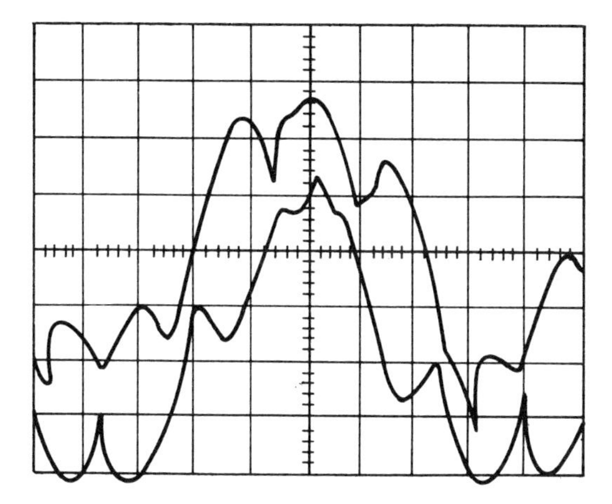

Figure A–7

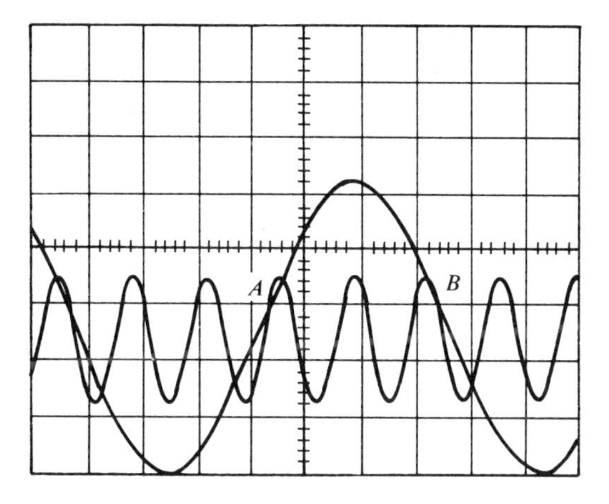

Figure A–8

Figures A–10 and A–11 illustrate the addition of the fundamental frequency plus the sixth (one-half cycle times six is equal to three cycles) harmonic. Note that at point *A* the harmonic wave is positive, and at point *B,* it is negative. Figure A–11 describes the effect of the additions and subtractions from *A* to *B*. Again, it is important to note that the resultant shows a higher peak at the left than at the right side of the curve.

Figures A–12 and A–13 illustrate the addition of the fundamental frequency plus the seventh (one-half cycle times seven is equal to three and one-half cycles) harmonic. Note that the harmonic wave at point *A* and *B*

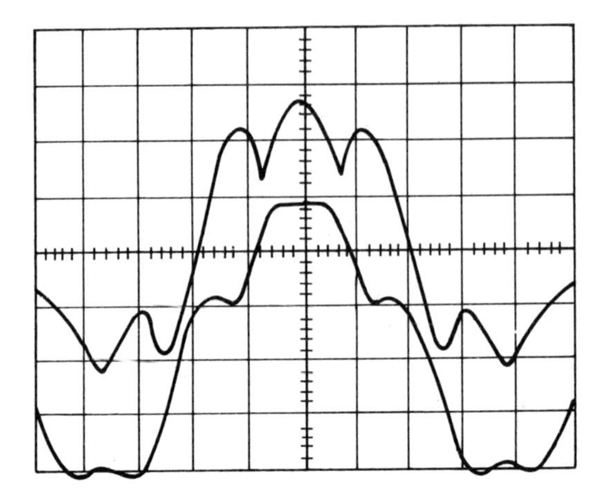

Figure A–9

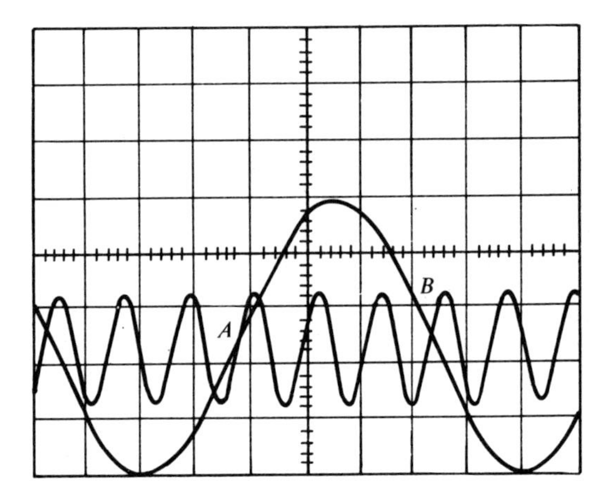

Figure A–10

is positive going and adds to the fundamental frequency, resulting in a waveshape of equal end peaks, with a larger peak at the center.

To summarize, it can be said that addition of even harmonics adds a tilt to the resultant wave shape. Adding odd harmonics tends to build up to a square wave. The resultant is a nonsinasoidal wave shape. The addition of all harmonics in a periodic wave shape results in a sawtooth shape (Figure A–14). The addition of all odd harmonics results in a periodic square wave (Figure 15).

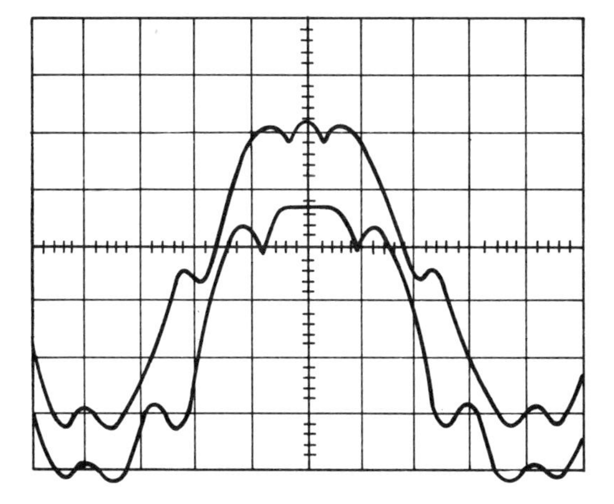

Figure A–11

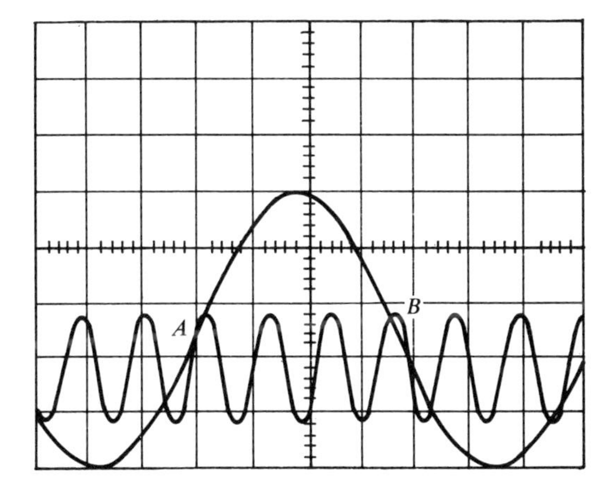

Figure A–12

To provide an output signal that is an exact reproduction of the input signal, the amplifying device must be capable of amplifying with uniform gain all of the components (harmonics) of the sawtooth and square-wave signals. This type of device is referred to as a wide-band amplifier.

Figures A–16 and A–17 show a comparison between a narrow-band amplifier and a wide-band amplifier.

The narrow-band amplifier in Figure A–16 is not capable of amplifying all of the harmonics, and produces an output wave form that is highly

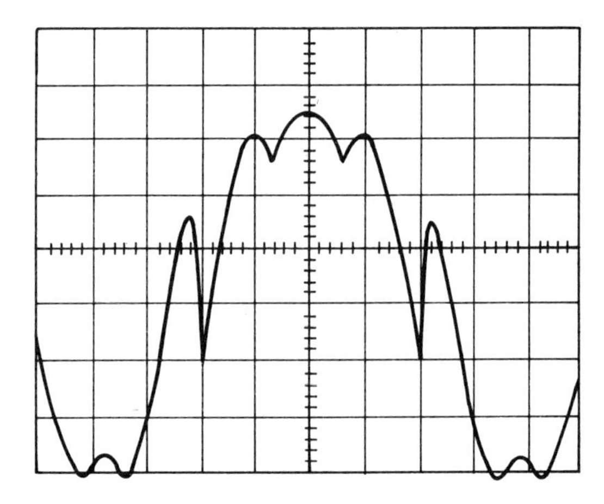

Figure A–13

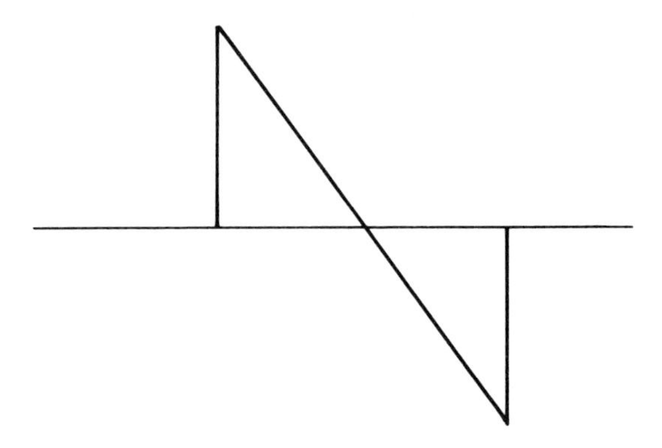

Figure A–14

distorted when compared to the input wave form. The wide-band amplifier (Figure A–17) can amplify all of the harmonics, and thus can produce an output wave form that is an exact reproduction of the input wave form.

Figure A–18 shows the effect of a wide-band amplifier on a square-wave input with the following defects:

1. Output with low-frequency attenuation
2. Output with excessive low-frequency gain
3. Output with phase distortion
4. Output with phase distortion

Figure A–18(e) shows the result of an amplifier with uniform gain and phase distortion.

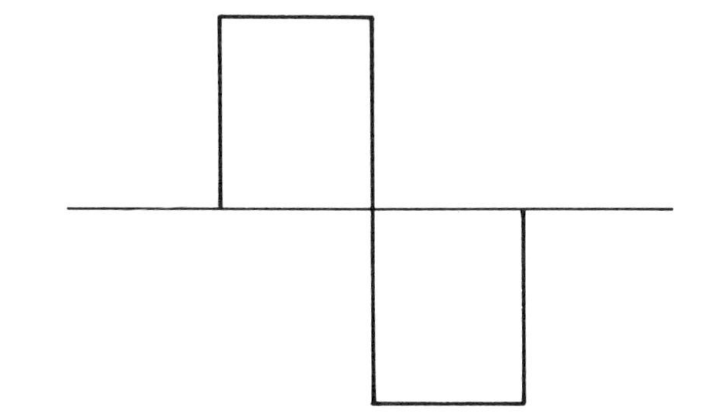

Figure A–15

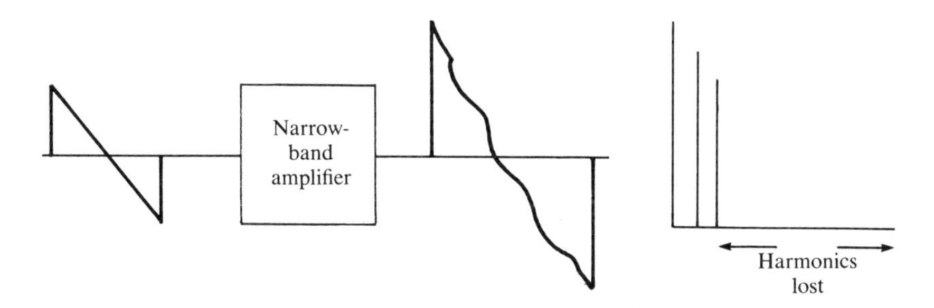

Figure A–16

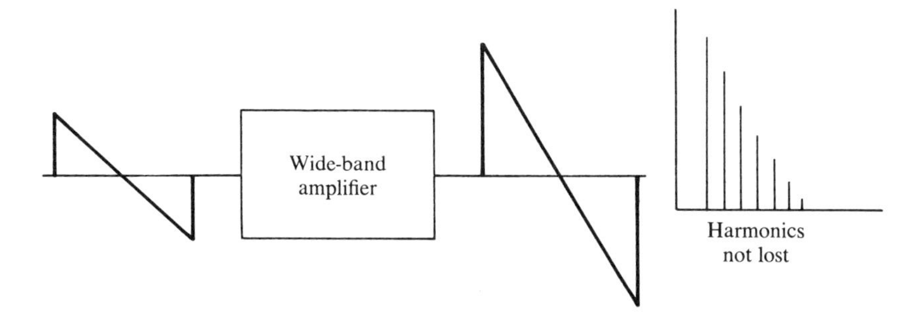

Figure A–17

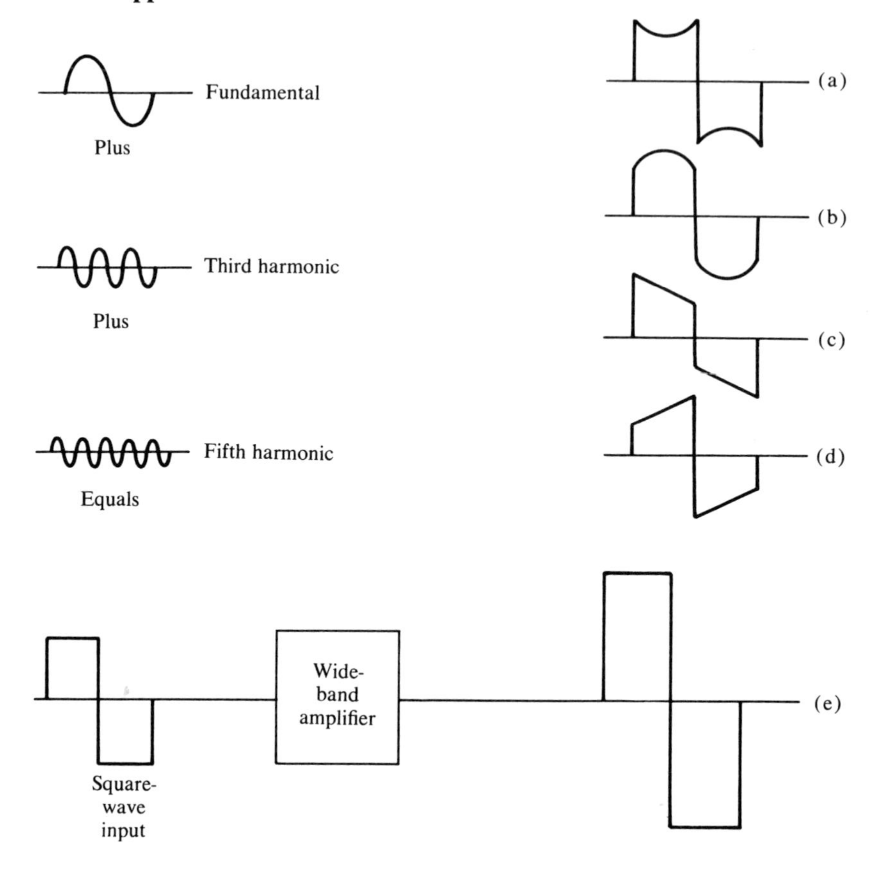
Fundamental
Plus
Third harmonic
Plus
Fifth harmonic
Equals
(a)
(b)
(c)
(d)
Square-
wave
input
Wide-
band
amplifier
(e)

Figure A–18

appendix B

suppression of color subcarrier in brightness signal

Figure B–1 shows a color signal going through a black-and-white receiver. The bandwidth response of the various stages is shown, but since the color information is low visibility information in the brightness channels, it is not to be seen on the black-and-white screen. The output of the video amplifier

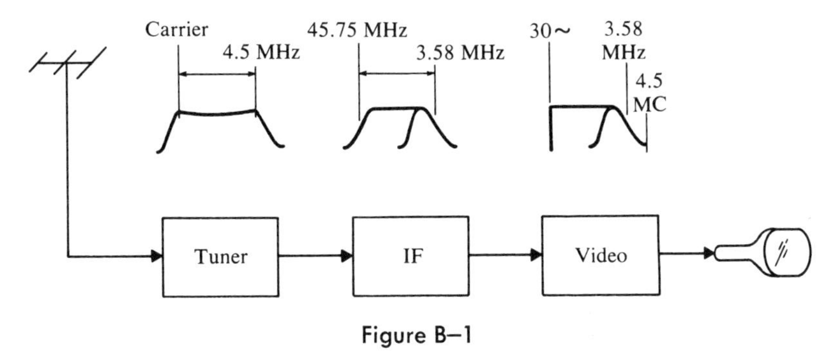

Figure B–1

will produce a signal on the cathode of the black-and-white picture tube, which represents the brightness level of the scene being transmitted in color. This will be a 100% signal for a white transmission, a 59% signal for a green transmission, a 30% signal for a red transmission, and an 11% signal for a blue transmission. The 3.58-MHz color subcarrier frequency was picked so that it would be a low visibility signal if received on a black-and-white television screen.

The relationship of the subcarrier to the horizontal-scanning frequency is an odd harmonic of one-half the horizontal-sweep frequency. This arrangement allows the 3.58 MHz to cancel itself on the television screen on successive scanning lines. Figure B–2 illustrates this.

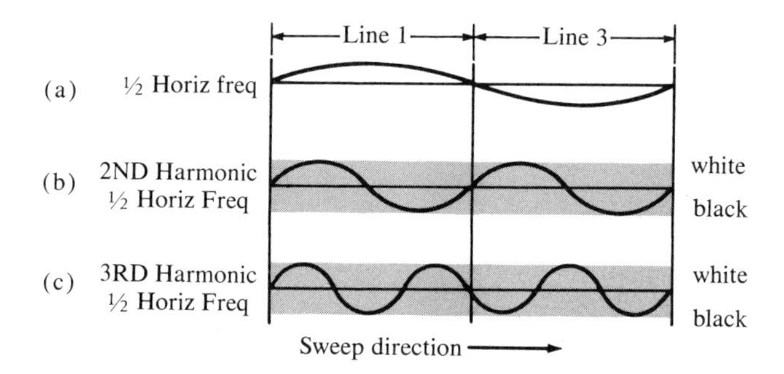

Figure B–2

Two successive horizontal-scanning lines, lines 1 and 3, are shown here. Since our television system uses a 2 : 1 interlaced method, the odd-numbered lines are scanned on one vertical field, and the even-numbered lines are scanned on the next field. Lines 1 and 3 are successive lines on one particular field.

In figure B–2(a), we show that a signal representing one-half the horizontal-sweep frequency ($1/2 \times 15{,}734.3 = 7{,}867$ Hz) would require two successive scans to complete one cycle. Figure B–2(b) shows that the second harmonic, 7,867 Hz ($2 \times 7{,}867 = 15{,}734$), would complete one full cycle for each scan line, and would add successive lines. This signal moves toward the white level at the start of each line, so it would appear as white on the left side of the television screen and as black on the right side of the screen. This condition would not be acceptable to the viewer.

Figure B–2(c) shows that the third harmonic (odd harmonic) of the 7,867 Hz makes the screen go white at the start of line 1, but dark at the start of line 3. This would cancel out on successive lines and not be visible. To visualize the above illustration, shift line 3 toward line 1 so that the sine waves are superimposed on top of each other. The cancellation process can now be seen.

The subcarrier frequency, 3.579485 MHz, is the 455 harmonic (odd) of one-half the line frequency ($455 \times 15{,}734/2 = 3.579485$). To obtain the correct effect, the horizontal-sweep frequency on color transmission has been changed from the black-and-white frequency (15,750 Hz) to 15,734 Hz. In order to keep the 262.5 : 1 relationship between the horizontal- and vertical-sweep rates, which is necessary to maintain stability in synchronizing, the vertical-sweep rate was changed from 60 Hz for black-and-white television to 59.94 Hz for color television. Both of these new rates are well within the tolerance of both black-and-white and color receivers.

The second condition exists in the color receiver, which practically eliminates the visual effects of any color subcarrier that could appear on the cathodes of the color picture tube.

appendix C

television channels and related information

Table C–1

Chan No	Video Carrier	Sound Carrier	Osc Freq	Video Image	Sound Image
vhf					
2	55.25	59.75	101	146.75	142.25
3	61.25	65.75	107	152.75	148.25
4	67.25	71.75	113	158.75	154.25
5	77.25	81.75	123	168.75	164.25
6	83.25	87.75	129	174.75	170.25
7	175.25	179.75	221	266.75	262.25
8	181.25	185.75	227	272.75	268.25
9	187.25	191.75	233	278.75	274.25
10	193.25	197.75	239	284.75	280.25

Table C–1, Cont.

Chan No	Video Carrier	Sound Carrier	Osc Freq	Video Image	Sound Image
vhf					
11	199.25	203.75	245	290.75	286.25
12	205.25	209.75	251	296.75	292.25
13	211.25	215.75	257	302.75	298.25
uhf					
14	471.25	475.75	517	562.75	558.25
15	477.25	481.75	523	568.75	564.25
16	483.25	487.75	529	574.75	570.25
17	489.25	493.75	535	580.75	576.25
18	495.25	499.75	541	586.75	582.25
19	501.25	505.75	547	592.75	588.25
20	507.25	511.75	553	598.75	594.25
21	513.25	517.75	559	604.75	600.25
22	519.25	523.75	565	610.75	606.25
23	525.25	529.75	571	616.75	612.25
24	531.25	535.75	577	622.75	618.25
25	537.25	541.75	583	628.75	624.25
26	543.25	547.75	589	634.75	630.25
27	549.25	553.75	595	640.75	636.25
28	555.25	559.75	601	646.75	642.25
29	561.25	565.75	607	652.75	648.25
30	567.25	571.75	613	658.75	654.25
31	573.25	577.75	619	664.75	660.25
32	579.25	583.75	625	670.75	666.25
33	585.25	589.75	631	676.75	672.25
34	591.25	595.75	637	682.75	678.25
35	597.25	601.75	643	688.75	684.25
36	603.25	607.75	649	694.75	690.25
37	609.25	613.75	655	700.75	696.25
38	615.25	619.75	661	706.75	702.25
39	621.25	625.75	667	712.75	708.25
40	627.25	631.75	673	718.75	714.25
41	633.25	637.75	679	724.75	720.25
42	639.25	643.75	685	730.75	726.25
43	645.25	649.75	691	736.75	732.25
44	651.25	655.75	697	742.75	738.25
45	657.25	661.75	703	748.75	744.25
46	663.25	667.75	709	754.75	750.25
47	669.25	673.75	715	760.75	756.25

Table C–1, Cont.

Chan No	Video Carrier	Sound Carrier	Osc Freq	Video Image	Sound Image
uhf					
48	675.25	679.75	721	766.75	762.25
49	681.25	685.75	727	772.75	768.25
50	687.25	691.75	733	778.75	774.25
51	693.25	697.75	739	784.75	780.25
52	699.25	703.75	745	790.75	786.25
53	705.25	709.75	751	796.75	792.25
54	711.25	715.75	757	802.75	798.25
55	717.25	721.75	763	808.75	804.25
56	723.25	727.75	769	814.75	810.25
57	729.25	733.75	775	820.75	816.25
58	735.25	739.75	781	826.75	822.25
59	741.25	745.75	787	832.75	828.25
60	747.25	751.75	793	838.75	834.25
61	753.25	757.75	799	844.75	840.25
62	759.25	763.75	805	850.75	846.25
63	765.25	769.75	811	856.75	852.25
64	771.25	775.75	817	862.75	858.25
65	777.25	781.75	823	868.75	864.25
66	783.25	787.75	829	874.75	870.25
67	789.25	793.75	835	880.75	876.25
68	795.25	799.75	841	886.75	882.25
69	801.25	805.75	847	892.75	888.25
70	807.25	811.75	853	898.75	894.25
71	813.25	817.75	859	904.75	900.25
72	819.25	823.75	865	910.75	906.25
73	825.25	829.75	871	916.75	912.25
74	831.25	835.75	877	922.75	918.25
75	837.25	841.75	883	928.75	922.25
76	843.25	847.75	889	934.75	928.25
77	849.25	853.75	895	940.75	934.25
78	855.25	859.75	901	946.75	940.25
79	861.25	865.75	907	952.75	946.25
80	867.25	871.75	913	958.75	952.25
81	873.25	877.75	919	964.75	958.25
82	879.25	883.75	925	970.75	964.25
83	885.25	889.75	931	976.75	970.25

index

index